John Spiropoulos

Small Scale Production of Lime for Building

A Publication of
Deutsches Zentrum für Entwicklungstechnologien – GATE
in: Deutsche Gesellschaft für Technische Zusammenarbeit (GTZ) GmbH

Friedr. Vieweg & Sohn Braunschweig/Wiesbaden

The Author:
John Spiropoulos is a graduate in the fields of building science and project management. He was instrumental in the establishment of a small scale building materials and minerals development company in Kanye, Botswana and was responsible for the project investigation and implementation of several building materials production projects including a small limeworks and a fired clay brickworks.

CIP-Kurztitelaufnahme der Deutschen Bibliothek

Spiropoulos, John:
Small scale production of lime for building :
a publ. of Dt. Zentrum für Entwicklungstechnologien – GATE in: Dt. Ges. für Techn. Zusammenarbeit (GTZ) GmbH / John Spiropoulos. – Braunschweig ;
Wiesbaden : Vieweg, 1985.
ISBN 3-528-02016-4

Published by Friedr. Vieweg & Sohn Verlagsgesellschaft mbH, Braunschweig
Printed in the Federal Republic of Germany by Lengericher Handelsdruckerei, Lengerich

ISBN 3-528-02016-4

Contents

Introduction

In many developing countries lime is produced in comparatively large, technically sophisticated limeworks, to supply a specific industry or group of industries, such as chemical or metallurgical works, or sugar refineries. In such cases the limeworks will be situated as close as possible to these factories. Where lime is produced to supply the building industry, the limeworks will probably be situated in, or close to, densely populated areas so that they may absorb as much of the output as possible. Secondary markets will be centres which are sufficiently close to the limeworks to be economical and which can be approached on roads of reasonable condition. Inevitably in such situations areas having a *low demand,* and which have *bad roads leading to them* or are a *long distance* from the limeworks are either left unsupplied or undersupplied at high prices. This often constitutes a large proportion of a developing country.

In these areas productive activities tend to be mainly traditional agriculture, sometimes only just above the subsistence level. Apart from work done in the fields the population is usually *unemployed* and consequently there is a lack of cash income. The industrial experience of the local population is limited and the *technical skills level is low*. In addition, it is usually difficult to recruit well trained technicians and managerial staff to work in remote areas and then to keep them there at a cost which can be supported by a small project. The risk of project failure in such areas is usually high. To reduce this, the limeworks must have as small an output as necessary to supply the *local demand*. It must be of a low technological level to enable the *use of locally available resources.* This assists in reducing capital costs to a minimum and also enables operation, and maintenance and repair work to be executed by the local population without the need for highly trained technicians or managerial effort. Further, in most cases it will be desirable to arrange that the most *labour intensive methods* are used to *maximize employment*. Care should be taken that employment does not conflict with the existing social organization. For instance working hours may have to be shared with labour in the fields. For these reasons small output, "low" technology solutions such as those described in this handbook are viewed as appropriate for the situation outlined above.

The job of the project planner, in general terms, is to systematically determine the various alternatives in design and methods of operation that could be used, and ascertain the implications of each, whether economic, social or physical. Each aspect of the project must be evaluated by weighing up the technological choices against their implication, and then decisions on each aspect must be viewed in the light of the project as a whole.

Since this type of project is generally of a high risk nature, special efforts must be made in the investigative and planning stage. Information must be collected and the project planned very carefully to minimize the chance of failure. The effect of failure is not only economic, it may also create disillusionment and frustration in the local population which could result in lack of interest in, if not resistance, to further industrial development efforts in the region. The project should be implemented, if possible, with a pilot phase which runs short of a full capital commitment, to test viability before going into full production and a full financial commitment.

The aim of this handbook is to provide a guide to the field worker who is involved in the project investigation, planning and implementation. It provides a set of procedures, technical and operating information, and ideas which, in the experience of the author, may prove to be of value. The handbook is designed to provide a frame of reference and serve as a stimulus for the development of solutions appropriate to the specific circumstances. It must be stressed that what is described in this handbook should not be adopted for use automatically without evaluation.

The handbook has been structured to emphasize the project investigation, planning and implementation aspects, since so much depends on the successful execution of these. Secondarily, technical and operating information is provided, focusing specially on the design, construction and operation of small (less than 15 tonnes per day) limeworks, using a small vertical shaft kiln of the Khadi Village Industries type. These have proved flexible and successful.

A conscious attempt has been made to provide a document which is as general as possible without detracting from its practical value. It is to provide ideas, yet be usable. Finally, a list of names and addresses of organizations which may be of assistance to the project worker has been included.

1. General Information

1.1 Geological description of limestone

Limestone is a sedimentary rock composed mainly of calcium and magnesium carbonate. It is formed by the deposition either of the skeletons of small creatures and/or plants (organic limestones), or by chemical precipitation, or by deposition of fragments of limestone rock, on the beds of seas and lakes. Limestones are contaminated to a greater or lesser extent by the deposition of sand or clay which are the source of the impurities found in them. Usually there is a difference in quality in a deposit from one layer to the next. The purest carbonates and the most suitable from the production point of view tend to be the thick bedded type. Carbonate deposits may be found in horizontal layers as deposited, or at an angle from the horizontal due to earth movements. They will vary in density, hardness and chemical purity.

1.2 Chemical composition of limestone

Limestone is made up of varying proportions of the following chemicals with calcium and magnesium carbonate being the two major components:

Calcium carbonate	$CaCO_3$
Magnesium carbonate	$MgCO_3$
Silica	SiO_2
Alumina	Al_2O_3
Iron oxide	Fe_2O_3
Sulphate	SO_3
Phosphorus	P_2O_5
Potash	K_2O
Soda	Na_2O

The two main impurities are silica and alumina with iron as the third.
For a general purpose lime, a limestone with an SiO_2 content of up to 3.5 % and AL_2O_3 content of up to 2.5 % may be used where purer stone is not available, whereas lime for building or road construction purposes may have an SiO_2 content of up to 10 % (perhaps slightly more) and an Al_2O_3 content of 5 %. An Al_2O_3 proportion of greater than 5 % will produce a semi-hydraulic or hydraulic lime.

1.3 Physical characteristics of limestone

The *colour* of most limestones is varying shades of grey and tan. The greyness is caused by the presence of carbonaceous impurities and the tan by the presence of iron.
It has been found that all limestones are crystalline but with varying crystal sizes, uniformity, and crystal arrangement. This results in stone with a corresponding variance in density and hardness (Boynton p. 21). For lime production purposes there are two factors related to limestones' crystallinity and crystal structure which are of specific interest:
Density or porosity is determined as the percentage of pore space in the stone's total volume. It ranges from 0.3 %–12 %. At the

lower end are the dense types (marble), and at the upper the more porous (chalk). Generally, the finer the crystal size, the higher the porosity but there are anomalies which suggest that each case be considered separately. A high porosity makes for a relatively faster rate of calcination and a more reactive quicklime.

Limestone varies in *hardness* from between 2 and 4 on Moh's scale with dolomitic lime being slightly harder than the high calcium varieties. Limestone is in most cases soft enough to be scratched with a knife. Marbles and travertines have the highest compressive *strength* whilst chalk has the lowest.

Due to the variance in porosity, the *bulk densities* of various limestones range from 2000 kg/m^3 for the more porous to 2800 kg/m^3 for the most dense.

The specific gravities of limestones range from 2.65–2.75 for high calcium limestones and 2.75–2.9 for dolomitic limestones. Chalk has a specific gravity of between 1.4 and 2.

1.4 Classification of limestone

High calcium limestone is composed primarily of the minerals calcite or Aragonite ($CaCO_3$) with a total oxide ($CaO + MgO$) content of over 95 %. It can be a fine to a coarse grained stone of varying porosity and hardness. *Chalk* is a soft, fine grained, highly porous limestone. The pure, white chalks can have up to 99 % calcium carbonate whereas the grey variety can have up to 20 % impurities i.e. only 80 % $CaCO_3$. *Dolomitic and magnesian limestones*, in addition to the $CaCO_3$, contain a relatively large proportion of $MgCO_3$. Usually limestones containing 20 % to 44 % $MgCO_3$ are referred to as „dolomite“ or as dolomitic limestone, and those containing between 5–20 % $MgCO_3$, as magnesian limestone. They both vary in purity, density, hardness and colour. *Marble* is a metamorphosed limestone. It is either high calcium or dolomitic and highly crystalline, dense and hard, and varies in purity. *Oolitic limestone* is a chemically precipitated limestone of high purity. *Travertine* is a hard limestone formed by chemical precipitation in hot springs. *Tufa* has the same derivation but is softer and more porous. *Hydraulic limestones*, sometimes referred to as argillaceous limestones have a high proportion of clay and silicia (15–20 %) and can be either high calcium or magnesian. When fired a lime is produced which can set under water (hydraulic lime).

There are several other types of limestones (less common and of limited value) such as coral limestones, shell limestones, marl, cherty limestones and stalagmites and stalactites.

1.5 Chemical reactions in the production of lime

Calcination

A. High calcium limestones:

$CaCO_3 + heat \rightarrow CaO + CO_2 \uparrow$

B. Magnesian limestones:

$CaMg(CO_3)_2 + heat \rightarrow CaCO_3 + MgO + CO_2 \uparrow$

(at around 750 °C)

$CaCO_3 + heat \rightarrow CaO + CO_2 \uparrow$

(at around 1100 °C)

Hydration

$CaO + H_2O \rightarrow Ca(OH)_2$

(water)

$MgO + H_2O \rightarrow Mg(OH)_2$

(water)

Simply stated, limestone plus heat produces quicklime and quicklime plus water produces lime hydrate.

1.6 Common uses of lime in developing countries

1.6.1 Use of lime in building construction

Lime is used in buildings in one of two forms, either as dry slaked lime powder or as putty. The choice of whether to use one or the other form depends on the preference of the builders, the availability of water at the production site and the available means of transport. There is little difference in terms of quality between the two.
The properties which make lime an excellent cementitious building material are the shape and fineness of the particles, which provide the plasticity necessary for a good workable mortar or plaster, and its chemical properties which are the mechanical, strength giving characteristics.
Historically, lime was used for building in such places as ancient China and Greece and, although since the discovery of portland cement its relative use has decreased. it is used in mortars and plasters in combination with cement in most countries of the world. Good mixing proportions are:

lime	cement	sand
1	1	6
1	2	9

The first mix is used where greater strength and resistance to weather are required and the second where these properties are unnecessary.
In situations where the sand to be used is very fine and/or not well graded, a mix of 1:1:7 can be used. Where the use of cement is limited in one way or another a small proportion of cement will suffice or none need be used at all. The use of such mortars or plasters is however limited since they are susceptible to weathering. They are best used on internal walls, external walls which are well protected from the rain, and in dry climates. The mix suitable under such circumstances is 1:2/3, i.e. 1 part lime to 2 or 3 parts sand. The same mix proportions are used for mortars and plasters. Special care must be taken where the lime is of low quality (relatively low hydroxide content) and the sand badly graded and fine.

Limewash as an external and sometimes internal wall coating is used in many countries today. Its advantages are that it is relatively cheap, easy to apply and gives a clean white finish, whilst its disadvantage is that it rubs off easily.
The durability of limewashes can be improved by adding small amounts of common salts or stearates. The recipes given by Bessey are:

"Quicklime	1 kg
Tallow	60 g
Water	2.5 litres" (p. 3)

Quicklime is broken into small lumps and tallow is placed in shreds over it. Enough water is then added to slake the quicklime. Once slaking is complete the remaining water is added and then the slurry is screened to remove any lumps and remaining shreds of tallow.
If only hydrated lime is available, powdered calcium stearate and water are added to it to produce a slurry. The recipe is:

"Hydrated lime	1 kg
powdered calcium stearate	50 g
Water as required"	

Lime in combination with pozzolanic material can be used as an alternative cementitious material to portland cement. This type of cement can be used successfully for most purposes other than structural concrete work, i.e. in plaster, mortar, production of building blocks, screeds etc. Some types of pozzolanic materials are: pulverized fuel ash, blastfurnace slag, volcanic ash, diatomatious earth, under-fired clay bricks, porcellanite and others. Each has a particular form in which it is most effective as a pozzolana.

1.6.2 Soil stabilization in road construction

The construction of roads in tropical and sub-tropical regions in areas where the soil is clay or laterite will require the use of a soil stabilizer. Lime, like portland cement, reacts chemically with certain constituents of the clay and acts as a binder. The effectiveness of lime as a soil stabilizer is increased in warm climates where its rate of strength development will be faster than in cooler areas. 2–5 % hydrated lime is mixed with soil from the road and spread over the road surface for compaction. The quantities required will depend on the compaction density, width and thickness of sub-base or base designed, and also on the proportion of available lime in the lime hydrate. For a road 11 m wide, sub-base 200 mm thick and a compaction density of 1500 kg/m^3, where 4 % lime is to be added, 130 tonnes lime per kilometer of road will be required.

For an existing limework with a low output (up to 15 tonnes), the implications of such a demand are:

– A long period of warning will be required to develop stockpiles, the development of which will require special financing arrangements since the stockpiling will cause excessive strain on cashflow.

– The accent will be placed on the quantity rather than the quality produced. It should be noted however, that the higher the quality the lower the quantities of lime required. The economic effects on transport cost, sales prices, etc. need to be carefully considered.

– A dry hydrate needs to be produced.

The possibility of installing a small limeworks near the road works, if possible in a situation where its operation will be continued after the completion of the road, must be considered as it may be more economic than transporting large quantities over long distances.

1.6.3 Lime in agriculture

Lime and limestone are used in agriculture to neutralize the acidity of the soil and to promote effective use of added fertilizers. Crushed limestone, dolomite, chalk, etc. are preferred as they are slower acting than lime and therefore last longer.

Other uses are:

- water purification,
- sugar refining,
- tanning,
- neutralizing acid mine water,
- oil well drilling,
- wire drawing,
- paper and pulp production,
- sewage treatment,
- metallurgical processing,
- petroleum refining,
- calcium silicate brick production,
- paint production

and others.

2. Project Investigation and Implementation

2.1 Schematic representation

Phase I	Phase II	Phase III
conception of idea and project initiation		
Socio-Economic Study		
	mineral resource investigation	
	site survey	
	market study	
	decision on technology and location of production site	
	production plan	
		economic feasibility calculation
		manpower plan
		financing
		marketing plan
		project implementation plan
		project implementation

2.2 Socio-economic study

At the outset of the project, before anything else is done, a well thought out study must be executed to direct the actions that follow. The proposed project should be viewed from an economic, social and an environmental point of view. The study should collect information on, and evaluate such matters as:

– population distribution and its composition,
– local productive activities and income generation distribution,
– employment patterns and limitations, e.g. such as would apply in a subsistence agricultural environment,
– traditional building methods and building materials utilization,
– availability and general level of quality of limestone deposits in the region,

– environmental constraints (legal and otherwise),
– national and regional development plans and legal matters affecting the proposed project.

A project which is not planned and implemented with this wide perspective runs the risk of either being economically unviable or not satisfying the social and environmental needs of the region, or both. The objectives of the study are to identify the socio-economic priorities and constraints of the region and hence establish provisionally:
a) the type and form of the lime products demanded and the prices that can be paid,
b) a sufficiently accurate estimate of the extent of distribution and nature of the demand, i.e. qualities, where required, consistency and certainty of demand, expected changes in the market and potential for development,
c) the preferred geographic location of the production site,
d) the limestone deposits that could possibly by exploited,
e) the type of technological level that is suitable.

It should be noted that although the nature and the extent of the study will vary in each situation, it should be kept to a minimum. Only information which is very clearly of value must be collected. Therefore, to make sure that extra and unnecessary work which, in addition to being a waste of time also costs money, is avoided, a clear statement of objectives must be prepared before commencing and adherred to as far as possible. Since the study will form a basis for project planning special note must be taken of national and regional plans and care taken not to conflict with their objectives. Most of the information will be available at the different ministries or other local institutions. Detailed studies are unlikely to be necessary.

2.3 Mineral resource investigations

To avoid the duplication of unnecessary investigative effort, all available geological surveys, mineralogical investigations, feasibility studies and other reports pertaining to the proposed project should be gathered for consideration. The following public and private organizations may have conducted studies which have been "filed forever", but which may be relevant: the Geological Survey Department, Roads and Rural Roads Dept., Building Dept., Dept. of Agriculture, Dept. of Mining, Dept. of Industries, as well as private organizations which either use lime in some form or have thought of undertaking this line of production.
Although the available documentation may be substantial, in most cases additional investigations will have to be conducted, to provide adequate information. The questions to be answered are:
1. What types of carbonates are available in the region of interest and where?
2. What qualities of carbonates are available and what is the consistency of the qualities of each deposit?
3. What quantities of acceptable, consistent quality carbonates are available?
4. What are the quarrying implications in each case?
The personnel and facilities for geological surveying and mineralogical testing of representative samples of material from the different deposits will normally be available in the local Geological Survey. If not, samples could be sent for testing outside of the country or simple field testing as described in section 4 may be sufficient.

2.4 Site surveys

The purpose of the site surveys is initially to provide sufficient information for the selection of the most suitable production site and subsequently to provide the information required for planning production. The extent of the survey will depend on the particular requirements but should provide the following information:

– physiscal conditions to ascertain the most favourable position from a technical and economic point of view, for the quarry and then the rest of the production site. Quarrying is the most expensive operation in the lime production process and therefore it should take priority;
– fuels available for firing the kiln;
– availability of machinery and equipment, materials for plant construction (e.g. fire bricks), spares and repairs and maintenance facilities;
– condition of transport and communication systems;
– availability of labour at the required skills levels;
– availability of water, electricity and fuels.

The site surveys should follow examination of the results of the geological, mineralogical and chemical tests of the various deposits prior to selecting the production location.

2.5 Market study

The objective of the market study is to test the producible lime product in the market to determine with sufficient accuracy:

a) saleable quantities,
b) prices (at least a price range),
c) continuity of demand,
d) extent of future developments.

This research will require a thorough door to door survey of major potential buyers such as the Building Dept., Rural Roads Dept., Dept. of Agriculture and Dept. of Mines and corresponding private sector users. It is almost certain that the quality of lime that can be produced using the methods to be proposed (mixed feed method), is not adequate for metallurgical and chemical purposes.

2.6 Location of production site and decision on technological level

The socio-economic study, the geological and mineralogical investigations, the site surveys, and the market studies will enable a decision to be made on the level of technology most suited to the circumstances and also on the most suitable production site location.

The choice of technological level will be made primarily on the basis of the quantities and qualities of the lime products required and the implications of the infrastructural and service conditions as described under Site surveys (2.4). Other factors affecting the choice of technology are the skills levels of the labour available, the distribution of the market and the local development policies relating to import substitution, employment creation and energy utilization.

The production location will be selected by carefully considering the physical, infrastructural and service constraints of the various potential production locations so as to:

a) minimize the distance of the site from the market,
b) maximize the economic effectiveness of the operation.

The site survey activities, the selection of production site and the choice of the most suitable technology overlap to a considerable extent and often during the investigations a strict step by step following of the process is not only difficult but a waste of time.

2.7 Production plan

The production plan must be formulated with the following criteria in mind:

– energy saving options take priority;
– limitation of low efficiency production methods and hence wasted labour time;

– limitation of the use of plant and equipment which could cause maintenance and repair problems. It is in the nature of machinery to break down at the most inopportune moments, and breakdowns, particularly if in remote areas often bring the operations to a standstill for long periods due to the long distances involved and possibly inadequate repair facilities. In short, if machinery cannot be easily and reliably maintained and repaired it should be avoided;

– limitation of the use of plant and machinery which requires highly skilled manpower for adequate operation. Skilled manpower is often hard to recruit and then to keep in the rural areas. Projects can become over-dependent on these highly skilled people.

In general, the guiding principle throughout the production planning stage is to design the operations to be as simple and cost effective as possible and further, to avoid conditions which would endanger the long term continuity of the project.

Besides designing the technical aspects of the project the following aspects of the production plan must also be formulated:

– site and plant layouts,
– flow of materials and manpower on site,
– stock sizes,
– labour content,
– industrial safety measures.

Production planning procedure

Decisions to be made	**Factors to consider**
hydration process design	
– method of slaking – product bagging or preparation for market	– sales level and required stock levels – quality of lime required – form of the product (i.e. putty, powder, slurry) – slaking time – constraints imposed by physical conditions – machinery and equipment available – maintenance and other services available – labour constraint
Kiln design	
– choice of fuel and firing technique – type of kiln – size of kiln – quality control methods	– output to correspond with rate of slaking and stock levels of slaked lime required – quality required (what variances are tolerable) – physical constraints – types of material available for construction – machinery and equipment available – repair and maintenance facilities – labour constraint – fuels available for firing

Decisions to be made	Factors to consider
Feed preparation	
– preparation method	– rate and type of feeding method to be used – type of carbonate to be used – machinery and equipment available – maintenance facilities available – physical constraints – labour constraints
Quarry design	
– quarry plan – quarry method	– physical conditions of quarry site – quantities of feed required – waste factor of feed preparation – type of carbonate of deposit – expected project life, alternative quarry methods and environmental considerations – machinery available – maintenance and repair facilities – labour constraint
Handling and transport	
– methods and equipment to be used	– physical conditions of terrain – cost of labour, productivity levels and employment creation policy – machinery and equipment available – repair and maintenance facilities – fuels available – labour constraint

2.8 Financing

Once the feasibility calculations are completed, and a decision has been made to proceed, formal application for project funding must be made. However, it is strongly recommended that the means of financing are investigated long before reaching this stage. The possible sources of finance should be identified and approached, firstly to determine their interest to invest in the type of industrial development planned and secondly, to establish the terms and conditions.

The type of financing organizations that may be approached, depending on the specific conditions are:

– central or local government development funds,
– banks and other local financial institutions,
– international or national development agencies,
– local private institutions.

There may be a considerable period between making an application for finance and it being received. Therefore, if the continuity of proceedings is important, that is, if a break between completing the feasibility study and commencing project implementation is undesirable, application for short term „bridging“ finance may have to be made. A commitment of investment from

the funding organization or another form of surety may be required to obtain this loan. The financing organization will require most of the following documents depending on the size of the capital outlay, the nature of the project and the associated risk:
– market study and plan,
– production plan,
– plant and equipment schedule including specifications and prices,
– schedule of buildings and other structures,
– capital cost schedule,
– production costs, cost of sales and overhead costs schedule,
– budgeted income statements (first year and five year),
– cost-volume-profit analysis,
– project implementation programme,
– manpower plan,
– cashflow, project breakeven date and working capital requirement,
– return on investment (ROI),
– payback period,
– discounted rate of return, social cost-benefit analysis, value-added, and perhaps some others.

2.9 Marketing plan

The purpose of the marketing plan is to define a course of action which will minimize the sales distribution and management effort and hence cost, whilst at the same time maximizing the satisfaction of the needs of the region. It should maximize the socio-economic benefit resulting from the undertaking, and ensure long term continuity and stability of the project.
The decisions that have to be made are:
– prices,
– areas of distribution,
– products of distribution, i.e. quantities and qualities,
– preparation of budget which indicates the amounts to be spent on advertising and promotional activities, selling, and distribution effort,
– project future developments.

2.10 Example economic feasibility calculation

The purpose of the example is to provide the project worker with an idea of the types of costs that could be incurred, calculations that may need to be carried out and a format to use to determine economic feasibility. The unit of currency has been omitted deliberately since this will vary from project to project. The figures are based on a specific case study of a small works in a "Third world" rural environment.
The example includes the calculations that are most likely to be required.
The capital cost schedule will be based on information collected in preparing the plant and equipment schedule, including their specifications and prices, and the schedule of buildings and other structures.

Capital cost schedule

plant and machinery

kiln – depreciated over 10 years	8000
loading hoist – depreciated over 5 years	1500
hydration plant – depreciated over 10 years	6000
water handpump and piping – depreciated over 10 years	4000
lime putty loading mechanism – depreciated over 5 years	2500
5 tonne tipper truck – depreciated over 5 years	20000
wheelbarrows, shovels, breaking hammers and buckets – replaced every 3 years	1800
safety clothing and equipment – replaced every 2 years	1000
office equipment and furnishings – depreciated over 5 years	3000
	47800

buildings

office, store and equipment building	12000
storage shed	1500
fencing	800
– all depreciated over 25 years	14300
	+ 47800
	62100
add 10 % contingency	6210
Total capital cost	68310

Implementation cost

(all costs to be incurred in the 6 month implementation period are included)

quarrying and dressing cost	3600
fuel cost	900
transport cost	400
labour cost	7460
superviser salaries	1920
rents and royalties	320
	14600

Total investment cost

total investment = total capital cost + project implementation cost
= 68310 + 14600
= 82910

Calculation of depreciation cost

For the purpose of this example let us assume that the capital costs are depreciated on a straight line basis, i.e. an equal amount each year for a number of years. Also, that the items after their period of use, will have no scrap value. (This is not always the case.)
The formula to use is:

$$\text{Depreciation} = \frac{\text{capital cost} - \text{scrap value}}{\text{number of years over which depreciated}}$$

Example:

$$\text{Depreciation of kiln} = \frac{8000 - 0}{10} = 800 \text{ per annum}$$

The calculation is carried out for each item on the capital cost schedule and the results are added to establish the final figure.

In this example the depreciation amount is 8812 per annum.

Production cost, cost of sales and overhead costs schedule

Production costs (per month)

cost of raw materials:	
– quarry and dressing cost	1200
– fuel costs	450
labour cost	1120
overhead cost:	
– supervisor's salary	320
– indirect materials	50
– repairs and maintenance	120
– replacement cost	80
– rents and royalties	100
– depreciation cost	740
– transport cost – administration	100
– blasting cost	200
– site administration costs (post, telephone, stationery etc.)	50
– water and electricity	30
– fuel oil	40
	4600

Cost of sales (per month)

marketing cost:	
– promotional work	130
– transport	150
– advertising	80
distribution cost (charged separately to buyer per load)	–
	360

Overhead cost (per month)

head office staff salaries	1240
loan repayment – paid over 10 years	642
office expenses	100
rent for offices	100
transport cost – administration	50
water and electricity	15
	2147

Notes:

– In this instance the production cost, cost of sales and overhead cost are calculated per month of normal production. However, it is not necessary for them to be calculated on a monthly basis. They should be calculated as is most convenient in the particular circumstances.

– These costs are calculated for purposes of preparing the annual and 5-year budget and are the result of careful compilation of information and costs for each item under each type of cost.

Budgets

Budgeted income statement – (year 1)

sales (1400 tonnes at 75 per tonne)	
Less	
cost of sales	4320
production cost	55200
overhead cost	25764
profit before tax	19716
less 40 % tax	7886
profit after tax	11830

Note:

– The first year's budget applies for the period commencing from after the implementation period, i.e. 7th month onwards.

– 40 % tax assumed.

Projected income statement (5 years)

	year 1	year 2	year 3	year 4	year 5
sales forecast	1400	1470	1545	1620	1700
volumes available for sale	1400	1470	1485	1485	1485
sales price	75	82	90	100	110
sales	105000	120540	133650	148500	163350
less					
cost of sales	4350	4752	5227	5750	6325
production cost	55200	60720	66792	73471	80818
overhead cost	9180	10098	11108	12219	13441
depreciation	8880	8880	8880	8880	8880
loan repayment	7704	7704	7704	7704	7704
profit before tax	19716	28286	33939	40476	46182
less 40 % tax	7886	11314	13576	16190	18473
profit after tax	11830	16972	20363	24286	27709

Notes:

– Demand for lime is estimated, from the market study, at 1400 tonnes for year 1 with an increase of 5 % per annum for the first 5 years.

– Output is planned at 1485 tonnes per year.

– The inflation rate is 10 % per annum with a corresponding increase in costs.

– Costs which do not alter from one year to the next such as in this case the loan repayment and depreciation, have been separated from the production and overhead costs.

– From the third year onwards it is estimated that the demand will exceed output. If these projections prove correct it is recommended that an extension of production be planned around the end of year 4.

Cost – volume – profit analysis

sales price (year 1)	75 per tonne
sales volume per month (year 1)	116 tonnes per month

variable cost – per month	
quarrying and dressing costs	1200
fuel costs	450
labour costs	1120
royalties	60
fuel oil	40
blasting cost	200
water and electricity	30
	3100

unit variable cost $\frac{3100}{116} = 26,72$

fixed costs – per month	
– supervisor's salary	320
– indirect materials	50
– repairs and maintenance	120
– replacement cost	80
– rents	20
– transport cost – production	100
– depreciation cost	740
– site administration costs	50
	1480
cost of sales	360
overhead cost	2147
total fixed cost	3987

Notes:

– The objective of this analysis is to determine the volume of sales necessary to cover fixed costs. It can be considered the minimum sales target and termed the Breakeven Sales Volume.

– The criterion for separating variable costs from fixed is that variable costs vary with changes in the level of production whereas fixed costs remain the same regardless of changes in level.

Sales price	75.00
Less: unit variable cost	26.72
contribution margin	48.28 per tonne of lime sold

(Each tonne of lime contributes 48.28 towards fixed costs.)

Fixed cost 3987 per month

Breakeven sales volume:

$$\frac{\text{fixed cost}}{\text{contribution margin per tonne sold}}$$

$$= \frac{3987}{48.28}$$

= 82.58 tonnes lime per month

Return On Investment (ROI):

$$\frac{\text{Return (year 1) (i.e. profit after tax, year 1)}}{\text{total investment cost}}$$

$$= \frac{11830}{82910}$$

= 14.27 %

Note:

The return to be used in the calculation is that which would be earned in a normal year, i.e. from the year commencing after the implementation period.

Payback period

year	Investment to recover (beginning) of year)	Income after tax	Investment to recover (end of year)
1	82910	11830	71080
2	71080	16972	54108
3	54108	20363	33745
4	33745	24286	9459
5	9459	27709	

Other calculations such as discounted rate of return, value-added and social cost-benefit analysis may be required, in which case specialist personnel may be consulted or the financing institution itself may be asked for assistance if the project manager is unable to calculate these figures.

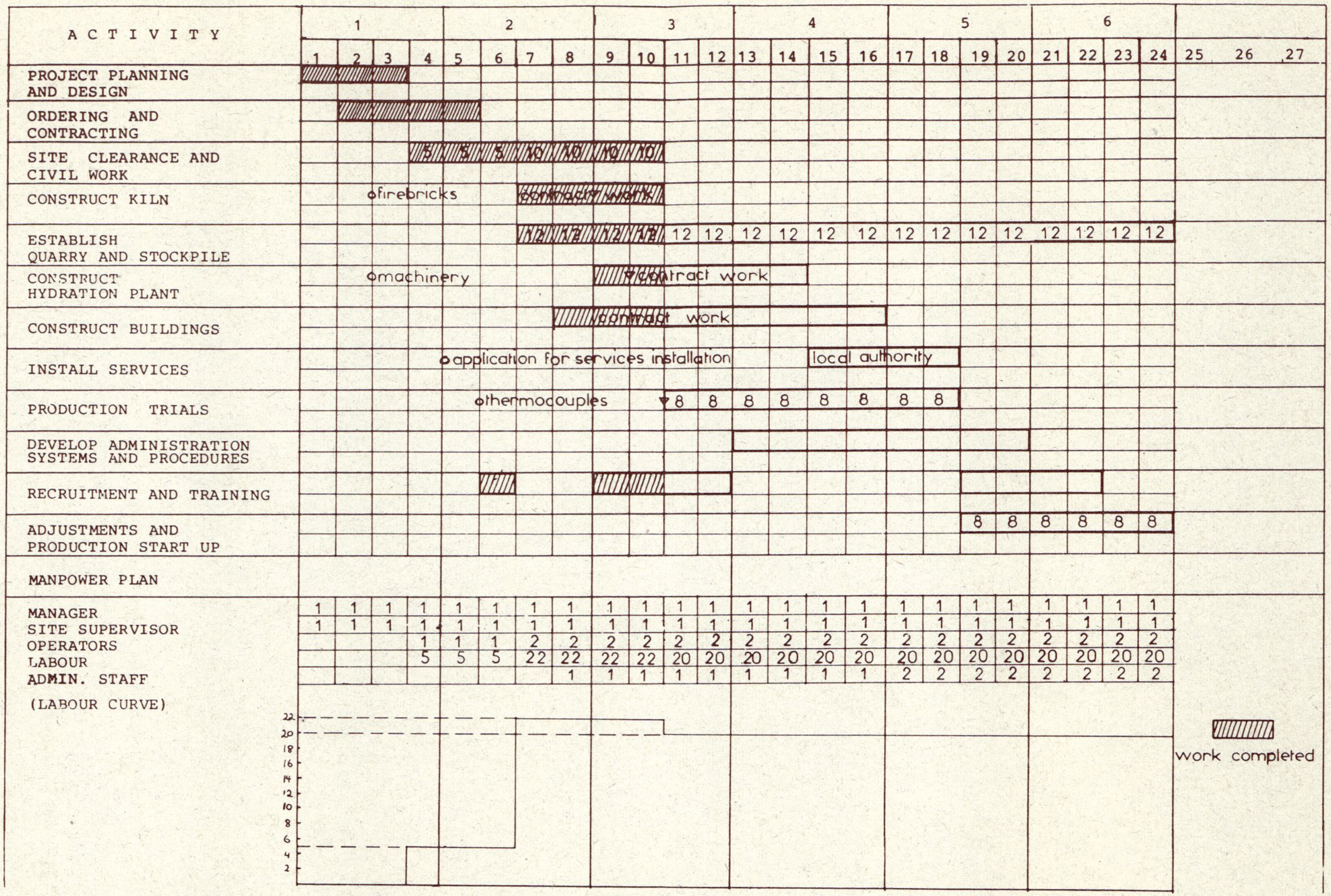

ACTIVITY	1	2	3	4	5	6	7	8	9	10	11	12	13	14	15	16	17	18	19	20	21	22	23	24	25	26	27
(months)	1				2				3				4				5				6						
PROJECT PLANNING AND DESIGN																											
ORDERING AND CONTRACTING																											
SITE CLEARANCE AND CIVIL WORK				5	5	5	10	10	10	10																	
CONSTRUCT KILN			firebricks				contract work																				
ESTABLISH QUARRY AND STOCKPILE							12	12	12	12	12	12	12	12	12	12	12	12	12	12	12	12	12	12			
CONSTRUCT HYDRATION PLANT			machinery						contract work																		
CONSTRUCT BUILDINGS								contract work																			
INSTALL SERVICES					application for services installation										local authority												
PRODUCTION TRIALS						thermocouples					8	8	8	8	8	8	8	8									
DEVELOP ADMINISTRATION SYSTEMS AND PROCEDURES																											
RECRUITMENT AND TRAINING																											
ADJUSTMENTS AND PRODUCTION START UP																			8	8	8	8	8	8			
MANPOWER PLAN																											
MANAGER	1	1	1	1	1	1	1	1	1	1	1	1	1	1	1	1	1	1	1	1	1	1	1	1			
SITE SUPERVISOR	1	1	1	1	1	1	1	1	1	1	1	1	1	1	1	1	1	1	1	1	1	1	1	1			
OPERATORS				1	1	1	2	2	2	2	2	2	2	2	2	2	2	2	2	2	2	2	2	2			
LABOUR				5	5	5	22	22	22	22	20	20	20	20	20	20	20	20	20	20	20	20	20	20			
ADMIN. STAFF								1	1	1	1	1	1	1	1	1	2	2	2	2	2	2	2	2			
(LABOUR CURVE)																											

work completed

Project Implementation Programme

2.11 Project implementation programme and resource plans

The implementation programme may be required by the financing organization but even if it is not, it is an essential management tool. It is not only used to plan the progress of the work, the use of resources and the procurement of machines, equipment and materials, but also to monitor progress during the course of implementation. The degree and form of its use can obviously best be judged by the project manager but it must be stressed that the programme's real value lies in its practical use. Particularly for this size and type of project a neat, "pretty" drawing stuck to the wall has limited value in that at best it describes past events only. It should be a sheet of paper on which notes are written, changes made and it should be figuratively, kept in the back pocket of the manager. It should serve to remind him or her about when to order materials, recruit personnel and also to clarify and concentrate his/her thoughts on the progress of the project (see example p.21).

The progress of the work is planned by plotting the duration of each activity along the time scale of the bar-chart. The sequence and duration of these activities will be finalized after a process of shifting them backwards and forwards along the time scale, increasing or decreasing the use of labour or any other resource so as to adjust the duration of the activities or, if necessary changing the method or process used in the activities. The resource plans necessary are the manpower and cashflow plans.

a) *Manpower plan* – The objective of the manpower plan is to smoothen the use of manpower thus minimizing the amount of hiring and firing and the associated cost and cashflow implications, as well as to set recruitment dates allowing suitable periods for induction and training.

The "smoothing" effort should be concentrated on unskilled labour since this portion of the manpower is the least costly and easiest to transfer from one activity to another.

The procedure for manpower smoothing is as follows:

1. Assign to each activity the number of labourers required per week of its duration.
2. Add up the number of labourers on each activity in each week of the project.
3. Draw the manpower curve.
4. Shift the activities backward or forward in time, so as to maintain as consistent manpower numbers as possible and to incur expenses when financially convenient.

b) *Cashflow* – The purpose of the cashflow is to plan the flow of cash into the project, i.e. grants, loans, earnings etc., and out i.e. expenses and capital purchases in such a way that cash is always available throughout the project implementation, interest payments are minimized and interest earnings maximized. By adjusting the sequence of the activities, shifting them along the time scale and, more important, by making suitable financing arrangements, the flow of money in and out of the project can be managed to the best economic advantage and with minimum risk (see example p. 23).

2.12 Project implementation

Once the implementation programme has been completed, feasibility assured and financing secured, it must be extended into a workable plan of action. The major activities on the bar-chart, such as the construction of the kiln or the quarry preparation, should be planned individually and broken down further into tasks which can be allocated to the various project staff. In a small project the

	1	2	3	4	5	6	7	8	9	10	11	12	13	14	15	16	17	18	MONTHS
INCOME																			
SALES VOLUME							46,4	69,6	92,8	116	116	116	116	116	116	116	116	116	
SALES							3480	5220	6960	8700	8700	8700	8700	8700	8700	8700	8700	8700	
DEBT								52910											
EQUITY (OWN FINANCE)	30000																		
BRIDGING FINANCE	60000																		
CASH INFLOW	90000						3480	58130	6960	8700	8700	8700	8700	8700	8700	8700	8700	8700	
EXPENSES																			
MATERIALS																			
-STONE FEED				1200	1200	1200	1200	1200	1200	1200	1200	1200	1200	1200	1200	1200	1200	1200	
-FUEL				300	150	450	450	450	450	450	450	450	450	450	450	450	450	450	
TRANSPORT COST-IMPLEMENT.			100	100	100	100													
LABOUR COST	165	1615	2320	1120	1120	1120	1120	1120	1120	1120	1120	1120	1120	1120	1120	1120	1120	1120	
SUPERVISOR SALARIES	320	320	320	320	320	320	320	320	320	320	320	320	320	320	320	320	320	320	
INDIRECT MATERIALS							50	50	50	50	50	50	50	50	50	50	50	50	
REPAIRS AND MAINTEN.									120	120	120	120	120	120	120	120	120	120	
REPLACEMENT COST												320							
RENTS AND ROYALTIES	20	20	40	60	80	100	100	100	100	100	100	100	100	100	100	100	100	100	
TRANSPORT COST-ADMIN.			50	50	100	100	100	100	100	100	100	100	100	100	100	100	100	100	
BLASTING COST		100	1200							1200						1200			
WATER AND ELECTRICITY					30	30	30	30	30	30	30	30	30	30	30	30	30	30	
FUEL OIL						240						240						240	
SITE ADMIN. COSTS				50	50	50	50	50	50	50	50	50	50	50	50	50	50	50	
COST OF SALES		360	360	360	360	360	360	360	360	360	360	360	360	360	360	360	360	360	
HQ-STAFF SALARIES	600	680	920	920	1000	1240	1240	1240	1240	1240	1240	1240	1240	1240	1240	1240	1240	1240	
OTHER OFFICE EXPENSES						265	265	265	265	265	265	265	265	265	265	265	265	265	
LOAN REPAYMENT										850	850	850	850	850	850	850	850	850	
BRIDGING FINANCE REPAYMENT								60000											
CAPITAL EXPENSES		13530	22880	28600	3300														
CASH OUTFLOW	1105	16625	28190	33080	7810	5575	5285	65285	5405	7455	6255	6815	6255	6255	6255	7775	6255	6495	
CASH FLOW	88895	72270	44080	11000	3190	(2385)	(4190)	(11345)	(9790)	(8545)	(6100)	(4215)	(1770)	675	3120	4045	6490	8695	

Project Cashflow – Year 1

Sales commende at 40% of full demand expected for the year in the 7th month and increase at a rate of 20% per month reaching the forecast level by the 10th month. Sales are given in tonnes. / Sales price is 75 per tonne for the 1st year commencing from the end of the 6 month implementation period. / The project is to be financed by debt (loan) and equity (own finance). A short term "bridging" loan will be arranged with the bank received in full at the beginning of the project and repaid fully at the end of the 8th month. / The loan repayment will commence 3 months after its receipt from the financial institution, i.e. from the 10th month. Interest on the loan is included in the amount repaid. / Blasting cost is incurred every 6 months. / Replacement cost estimated at 80 per month and is incurred every 4 months on average. / Fuel oil is bought in bulk every 6 months. / Capital expenses are incurred in the implementation period and normally payment will be made for a time after this period. In this example, however, it is assumed that payments are made immediately. / Cashflow for each month is calculated as follows: Previous month's cashflow, inflow, outflow. / The *Working Capital Requirement* is equal to the highest negative cashflow shown, i.e. the highest figure in brackets in the *Cashflow* line. The working capital is required at the end of the 5th month to cover all the negative cashflows of the following months. The amount required is 11 345 (see month 8). / The total financing required is therefore part of the capital expenses and the implementation cost (52 910), and the working capital required (11 345), i.e. in total 64 255. / Paid over 10 years, the amount to be repayed annually is 6 425–50 with say, a payment of 1285 annually for interest on loan, totalling 7 710–50. The monthly loan repayment will be 642. / The bridging finance amount, bridging finance repayment, loan amount to be received in the 8th month, and loan repayment should be adjusted in the above cashflow to establish the *cashflow plan* which can be used to monitor the cash situation on a monthly basis.

manager and possibly an assistant may carry out all the tasks. The process of breaking down the activities into tasks can in any event be recommended as this will assist in clarifying the implications and consequences of a certain course of action. Task completion dates must be set and, if other project staff are involved, the project manager must arrange matters so that he remains informed on progress. Most important and something which should be prepared at the outset of implementation, are order and delivery dates of purchases critical to the progress of the work. Similiarly, the dates of request for, and arrival of, specialized personnel, contractors and subcontractors should be established.

The *detailed engineering design and specifications* could either be prepared by the project manager or specialized contract personnel. There are various organizations which might be of help for this activity (see Appendix I).

Procurement is the work involved in bringing a piece of machinery, equipment or material to the site. It involves the following work:

– Identifying suppliers, requesting quotations and an approximate date of delivery. For major items quotations will have been received during the stage when the capital cost was being determined.
– Selecting suppliers,
– Ordering, including the arrangement of import licences and customs clearance if necessary, and transport,
– Agreeing on place and date of delivery, and handling to place plant in position ready for installation.

The project manager must arrange the *contracting out of building and construction work, and plant installation.* It may be necessary under certain circumstances to hire labour on contract and do the supervision personally. If this is the case it will probably be necessary to hire project administrative staff, e.g. bookkeepers and clerks, at this early stage of the project. The work involved in contracting is as follows:
– preparing tendering documents, i.e. drawings and specifications,
– putting work out to tender,
– selecting tender,
– preparing contract documents and setting starting and completion dates.

If one is managing the construction work personally, preparation must be made for this work. It is recommended that a competent supervisor be selected who will concentrate on recruitment of tradesmen and labourers, liaise with the project manager on material requirements and technical matters, and supervise the work. If this supervisor proves suitable, he can be retained as supervisor of the limeyard after implementation. This should be borne in mind when selection is carried out.

The tasks involved in *site establishment and plant erection* are:
– clearing of access roads,
– clearing of site in preparation for construction and erection work,
– clearing of access to quarry site and clearing ground cover at quarry,
– arranging for installation of services at production site, i.e. water, electricity, telephone and toilets,
– constructing buildings,
– installing machinery and equipment for quarry, kiln and hydration plant.

Production trials must be carried out to determine necessary adjustments; see section 5 on production trials in Botswana. *Labour* must be recruited and trained.

A review of economic calculations should be prepared so as to make certain of the project viability and to review possible effects of adjustments resulting from the production trials before the project goes into full production.

3. Technical and Production Information

3.1 Quarrying and kiln feed preparation

Economic quarrying and preparation of stone for firing is a vitally important precondition for the long term viability of the project. It can easily be underrated, often resulting in wasteful, uneconomic practices which endanger the viability of the project. The techniques and methods adopted will depend on the particular conditions at each location and therefore nothing specific can be advised but it must be stressed that careful planning is essential. This stage of the production process must be planned with the following aims in mind:

a) to provide safe working conditions,

b) to install as economic a process as possible without eliminating available working places unnecessarily and without making the working conditions unnecessarily difficult,

c) Specially trained and experienced personnel will be required to carry out a geological survey, advise on the quarry plan and also to do blasting if necessary.

3.1.1 Selection of site

As a general rule, the most convenient quarrying location for a low output limeworks is one where stone is quarried from the side of a hill. The stone would be prepared (or dressed) at the quarry face and taken down the hill directly into the kiln.

The following factors should be taken into consideration when selecting a suitable quarrying site.

Geological factors: Quality of stone and the ease of extraction are the criteria for quarry site selection.

If a choice of sites exists they must be set off against each other, and the stone with the best quality at the lowest extraction cost selected. The ease of extraction will depend on the type of material to be quarried and the geological setting in which it is found.

Often there is a substantial difference between the quality of stone in one stratum and that in another. To acquire a uniform kiln feed, the preferred strata should be worked separately either leaving the waste material behind or removing it to a waste dump. However, if the difference in quality is small the different strata should be well mixed before being fired.

The ideal stratification is one where the dip of the bedding is sloping slightly up into a hill and where the usable layer/s are thick and easily accessible. The downward sloping (down into the hill) quarry makes the transportation of the quarried stone up the slope and to the kiln more difficult and will generate drainage problems in the rainy season.

Further, it should be noted that it may be easier and cheaper to quarry the strata along the side of the hill rather than into it, especially if the strata dip downward. Also, quarrying should follow the dip or strike rather than cut across it, which would result in a greater variance in feed quality.

Overburden: The thickness and hardness of the overburden, and the corresponding cost of removing it, are the other important

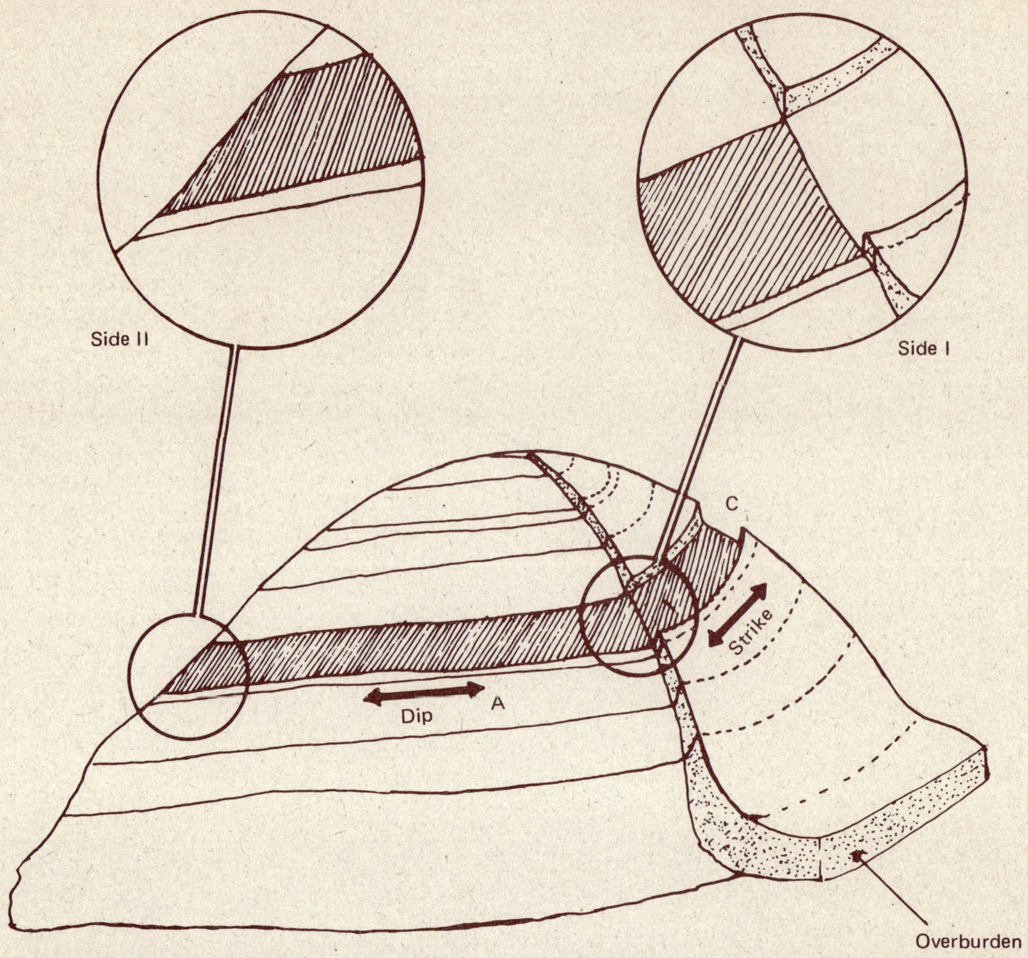

Hillside Stratification and Quarrying Alternatives

Note: Quarrying along the dip A would make for a quarry floor which is too steep. To avoid drainage problems during the rainy season and the additional effort required to haul the quarried stone up the slope of the quarry floor and then down the hill to the kiln, the area C would have to be excavated as one progressed into the hill. The situation on side II of the hill is preferable. It would be advantageous in the case described above to quarry along the strike as far as possible before quarrying into the dip.

factors for determing the acceptability of a lime deposit.

A thick or difficult to strip overburden could easily render a small project unfeasible due to the relatively high cost of stripping manually. It may very well be cheaper and will certainly be quicker to obtain the assistance of the local division of the Roads or Public Works Department, or even to hire a private roads contractor, to strip and remove the overburden, especially if additional access road and site clearance work needs to be done anyway.

Deposits on a hillside are likely to have less overburden than those on level ground but they may be more difficult to strip. It is advisable to dispose of waste a reasonable distance away from the quarry to avoid having to move it again in the future.

3.1.2 Quarry plan

The quarry plan consists of decisions on:

a) position and layout of the quarry,
b) flow of material, i.e. haulage, stock piling,
c) the extraction techniques, i.e. blasting method and plant equipment to be used,
d) dressing techniques,
e) future developments of the quarry.

Since each situation presents its own conditions, only general recommendations are made:
– The quarry layout should be such that it allows for easy access and manoeuvreability of loading and haulage machinery or animal driven carts on the quarry floor. At the same time, it should accommodate comfortably all the necessary manpower and their movements in the quarry. Double handling should be avoided.
– The choice of whether to use manual or mechanical means of dressing, loading or transporting depends on the implications of the alternatives in the particular context, but in general in a low output works, in the rural areas of a developing country, manual means are likely to be the most suitable.
– The method of extraction selected should be such that it ensures an adequate supply of stone feed with a limited amount of dressing effort. For example if blasting is necessary, the techniques used should be such as to limit the need for secondary blasting and provide a maximum amount of stone of the required size.
– Care should be taken to avoid flooding during the rainy season. It is advisable to acquire specialist knowledge for this as well as for the quarry plan in general. The local Geological Survey and Mining Department (government) may have a section with personnel to assist small scale miners.
– The future physical development of the quarry must be considered.

3.1.3 Quarrying

Stripping is the removal of overburden such as sand, gravel or clay of varying densities in preparation for quarrying. The ease with which it can be removed and the economic implications of the means available will determine the means used.
In a small project where the amount of soil to be removed is small and also relatively loose, the overburden can be stripped manually, using pick and shovel and wheelbarrows. An animal driven plough could be used to loosen the soil if it is hard.
If mechanical means such as bulldozers, scrapers and trucks are available that may be preferable both in terms of economy and efficiency.

Extracting is the removal of limestone from the quarry face. It should be carried out in a safe and systematic manner to provide a size of stone which will reduce the amount of effort required in its preparation for firing. The methods used, depend mainly on the nature of the material. The softer type limestones could be extracted by means of chisels, wedges, crowbars and picks and shovels, or by compressed air hammers, with or without a light blast depending on the specific conditions. The harder variety how-

ever, such as marble or dolomite, will require blasting. The type of blasting used depends on:

a) the nature of the joints and fissures in the rock, i.e. number, size and relative position of joints in the rock;
b) size of material required;
c) explosives available and relative costs;
d) drilling equipment available and the ease with which it can be used, i.e. hardness of material and topography of quarry.

It is essential that blasting be carried out by specialists. Blasting is a skill acquired through many years of experience and if conducted by "amateurs" can be both unsafe and uneconomic.

Dressing constitutes the preparation of stone, after it has been extracted from the quarry face, to the required size and shape. The methods used can be either mechanical or labour intensive.

The labour intensive methods are the use of hammers and chisels of varying sizes. A system of dropping a heavy metal ball on large pieces of rock to break them could also possibly be used. The most important thing to note when using labour intensive methods is that stone is easier to break along the grain than across it.

Mechanical methods of dressing and handling may become feasible when a small limeworks and a stone crushing operation are going to be implemented together. A saving in the blasting cost will be achieved since all the stone blasted will be used. If a stone crusher is not implemented concurrently, it could be that up to 30 % of the stone quarried would be lost in dressing. Further, the demand for stone by a small limeworks alone will not occupy a mechanical plant to an extent which will make its operation economic. The same applies to highly mechanized loading and haulage machinery.

Transportation: The dressed stone can be transported by wheelbarrows, wagons or tubs on rail (of wood or metal), animal driven carts, or by aerial ropeway.

Stock piles: If the cash availability permits, it will be cheaper when contracting for the blasting, to blast stone for a 3 to 6 month demand rather than having the blasting contractor transport equipment out to the site more often, especially when the distance from the contractor's site to the limeworks is great. Furthermore under conditions where there is a possibility of plant breakdown or labour availability fluctuation, for example where people employed in dressing may need to work in the fields for part of the day or for a period, it may be necessary to separate the stone preparation from the limeburning operation. This can be done by increasing the stock pile of kiln feed.

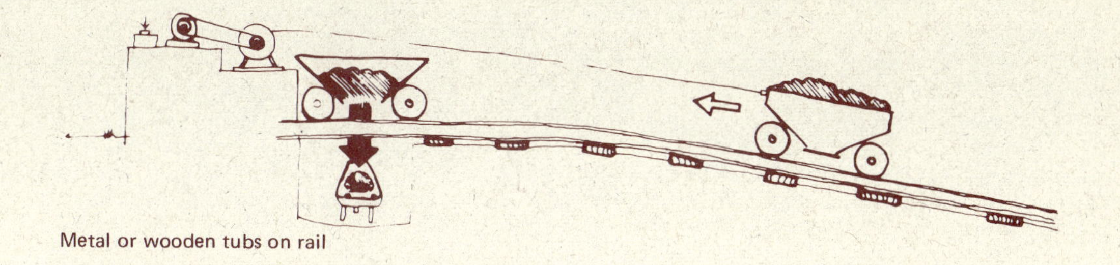

Metal or wooden tubs on rail

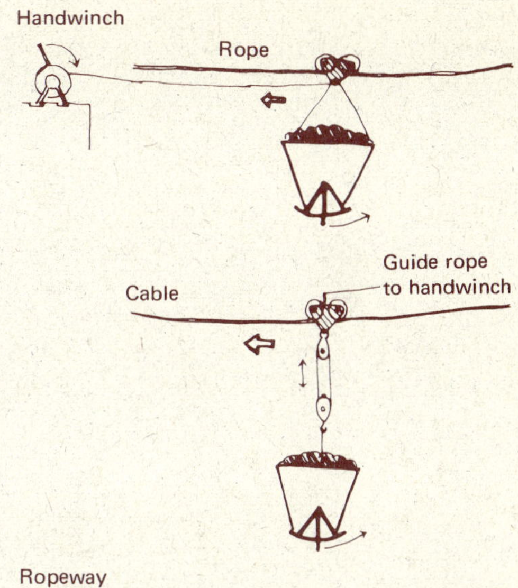

Ropeway

Transport and Handling Methods

Kiln Loaded by Skip on Rails

3.2 Fuels

3.2.1 Heat requirements and thermal efficiency

Calorific values

coals and cokes		
25120	–	37680 kJ/kg fuel
fuel oil		
37680	–	41870 kJ/kg fuel
wood		
12560	–	14650 kJ/kg fuel
producer gas (coal)		
348	–	855 kJ/m^3

The theoretical heat requirement for burning high calcium lime is given by Boynton as 770 kcal/kg (i.e. 3.2 million kJ/tonne). This is a minimum heat requirement. In practice it is inevitably exceeded due to heat losses. Heat is lost through:

- radiation,
- products of combustion,
- carbon dioxide evolved,
- discharged quicklime,
- drying stone,
- dust from kiln.

The heat required in practice depends on the "thermal efficiency" of the operation which in turn depends on the nature of the limestone fired, the type of kiln, nature of the fuel used and the skill of the operator.
Thermal efficiency is expressed as:

$$\text{thermal efficiency} = \frac{\text{3200000 kJ per tonne limestone} \times \text{\% available oxide content}}{\text{calorific value per kg fuel used (in kJ)} \times \text{kg fuel used per tonne lime}}$$

According to Bessey, thermal efficiency can vary between 30–85 % in large modern kilns, and be as low as 20 % in simple wood-fired batch kilns.
Assuming a well designed and operated simple mixed feed vertical shaft kiln has a thermal efficiency of 40 %, the fuel to be used has a calorific value of 30000 kJ/kg and the % available oxide is 85 %, the amount of fuel required per tonne of limestone charge would be:

$$40\,\% = \frac{3200000 \times 85\,\%}{30000 \times \text{fuel required}}$$

$$\text{fuel required} = \frac{320 \times 85}{3 \times 40}$$

$$= 227 \text{ kg fuel/tonne limestone}$$

Searle says "Vertical kilns differ greatly in their efficiency, and fuel consumptions of 400–640 lbs of coal or coke per ton of lime are common" (p. 391), i.e. approximately 182–290 kg per tonne limestone.
In a mixed feed operation the mixes of 1:4 coal to stone, 2:3 wood to stone and 1:2 charcoal to stone, are weight ratios which could be adopted to commence trials with, to be adjusted later as required. If the calorific value per kilogram fuel and the % available lime is known, a more accurate estimate of quantities can be made using the method described above. Probably this will change with experience in firing.

3.2.2 Fuel types

The selection of fuel, besides the kiln design and operation implications, depends on the cost per thermal unit (kJ), the behaviour of the fuel in use and the quantity and nature of impurities. There are various types of fuel that can be used for small scale lime burning; coal, coke, anthracite, lignite, peat, wood, as well as producer gas and fuel oil.

Coal differs in calorific value, in the degree of volatility and the type and quantity of impurities. Coal having a moderate to low volatile content is best for use in mixed feed operations, i.e. bituminous to subanthracite coals. The use of high volatile coals in such operations will result in part of the heat value escaping with the exhaust gases. They are best used in gas producers and to a lesser extent in external fireplaces.

The major impurities and levels of acceptability are:

– Volatile sulphur which burns to sulphur dioxide and is absorbed by the lime should be less than 2.5 %.

– Ash content should be less than 6 % but could go up to as much as 10 %. Low calorific fuels tend to have a higher proportion of ash.

– Water content should not exceed 6 %. A high water content results in excessive cooling of the flame and therefore a waste of heat.

The size of the coal fired should be as uniform as possible and at least in the size range of 10–40 mm diameter.

Coke is a very good fuel for limeburning either mixed with the limestone or to produce gas. A pure lime can be produced in a mixed feed operation by using a furnace coke since it contains a small amount of ash and sulphur. It is, however, difficult to light and burns less easily. For this reason, soft burnt coke produced in coke ovens is preferable. Gasworks coke is not satisfactory since it contains a high level of sulphur.

Anthracite can be considered as the best fuel for a mixed feed operation, possibly with the exception of wood. It is a coal with very little volatile matter. It has a high calorific value and ignites more easily than coke. The only likely disadvantage is its high cost.

Peat is a fuel with a low calorific value and high volatility. It would be best used to produce gas but could be used dry in a mixed feed operation to produce agricultural lime.

Lignite/Brown coal has the same low calorific value and high proportion of volatile matter as peat but also contains a large proportion of water. It is best used to produce gas. The producer should be larger than one for coal since lignite is a bulkier material. In addition, provision must be made in its design to cool the gas to condense the excess water so that it may be removed. Lignite often occurs in large quantities in locations close to limestone and therefore may be a more economic alternative to coal transported over a long distance.

Heavy *fuel oil* or a mixture of heavy oil and used motor oil can be used to produce a very good quality lime. A simple method is to transform the oil into an oil vapour which is then mixed with a suitable proportion of air and ignited in chambers around the periphery of the kiln to produce a flame which is fully developed before it leaves the combustion chamber, i. e. before it comes into contact with the limestone. There are various methods of atomizing the oil and mixing it with the necessary amount of air. In principle, the oil is preheated to a runny consistency and then injected under a slight pressure into a chamber where a blast of steam or air under pressure vaporizes it. The oil vapour and air mixture is ignited and the flame produced is directed towards the column of limestone in the kiln. The skill of the operator lies in his ability to mix the necessary amount of air and oil, or steam and oil, so as to produce a flame of such a length and intensity as to fire the limestone column to the centre without overburning any part of it, and without choking up the pores of the limestone with soot due to too lean a mixture.

In modern applications the oil is injected into specially designed gasification chambers where it is mixed with a predetermined mixture of air and recirculated hot flue gas, and oxidized to produce hot fuel gas which is introduced into the calcining zone where it ignites. These methods are generally more efficient but are comparatively expensive to implement and generally inappropriate for small scale production.

The oil must have a sulphur content of less than 2.5 %.

The type of *gas* to be used in a small scale application is producer gas. This is a good fuel for burning lime with the advantage of providing the possibility of using low grade fuels which could not be used in a mixed feed operation. The disadvantages are that its use requires a higher capital outlay than a mixed feed kiln and also more skill and effort in operation. If used correctly it can produce a flame similar in quality to that of wood with the advantage of not contaminating the lime. Further, a more evenly burnt product could be obtained since it facilitates better control of firing. Producer gas is made by passing a mixture of steam and air through a bed of fuel at least 1.5 m deep. The specific design of gas producer will depend on the type of fuel used. For example lignite and wood will require a deeper bed than coal. The fuel is charged at the top of the column and as it descends it is dried by the rising hot gases. It descends further in the column and acquires the temperature at which it parts with tar. Further down the column the fuel is decomposed into a gas mixture of carbon monoxide and some hydrogen. Lower down it is burnt in the ordinary manner providing the heat necessary for producing the gas in the upper layer. The ash falls through a grate into a pit from which it can be removed manually.

The gas can be produced by passing air through the bed of fuel alone but the use of steam together with air is preferable. Steam will prevent clinker forming on the grate and also produce a cooler flame. The consistent and uniform supply of air and steam is of the utmost importance for the production of a good and consistent supply of gas. (See p. 43, producer gas fired kilns.)

Wood is the traditional fuel for limeburning and from a technical point of view is ideal. It produces a long flame which is able to penetrate further into the limestone mass and throws off heat slowly and uniformly promoting even firing. In addition wood, on being fired, generates a considerable amount of steam which tempers the flame temperature. It is therefore almost impossible to overburn any of the lime when using wood as a fuel.

The use of wood is particularly effective in batch kilns. Dry or green wood can be used, preferably of the hard, dense varieties. It should be cut into lengths of approximately 250 mm if it is to be used in a mixed feed kiln, and if in an updraught type to lengths which suit the particular requirements.

The use of wood needs very special attention to avoid denuding the landscape which, in addition to the less important aesthetic damage, could cause soil erosion problems and may limit its availability for other uses, e.g. in rural areas of developing countries for heating and cooking purposes. In principle, if it is used it should be replaced.

Wood chips and other combustibles could be burnt in specially designed external fireplaces. Biogas is another form of fuel that could possibly be used. Theoretically, many types of fuels could be used but the practical implications often limit the range of choice. It is essential to ensure that an adequate and dependable supply exists in the long term.

3.3 Limeburning

The purpose of this section is to provide the project worker with:

a) a basic understanding of the principles of limeburning (calcination) and the factors which determine successful production of lime;

b) a general description of kilns that can be used in small scale limeburning;

c) technical and operational information on small vertical shaft mixed feed kilns.

3.3.1 Principles of limeburning and factors effecting the production of lime

3.3.1.1 Physical factors

Stone size: – A small stone can be fully calcined more easily and quickly than a large one. The selection of stone size will be based on the necessary temperature of firing and the corresponding firing time. These will depend on the physical and chemical characteristics of the stone.

For a given stone type, the size of the stone can be increased or decreased affecting the draught through the kiln and thus the rate at which the heat rises up the kiln shaft. The rate at which the fire rises up the shaft determines the length of time a stone is fired at the calcining temperature (firing time). Large stones have larger air gaps between them allowing air to flow through the shaft easily. Smaller diameter stones however, make the flow of air through the kiln difficult resulting in a slow rising of the fire and possibly overburning. Therefore, if a small stone is used the draught should be increased, by lengthening the kiln shaft or chimney, to avoid overburning.

For a shaft length of around 4.5 m and a chimney height of 1.5 m the following approximate stone sizes would probably be satisfactory:

– coarse crystalline	100 mm diameter
– fine grained crystalline marble or dolomite	150 mm diameter
– poorly crystalline	175 mm diameter
– porous (chalk or marl)	200 mm diameter

Stone shape: – It is preferable to use stone of a cubical or spherical shape. Flaky material will tend to obstruct the flow of air through the kiln, causing channelling and a bad distribution of heat.

Grading of stone: – A uniform, consistent size of feed is of primary importance. A variance in stone feed size will result in the smaller size being overburnt or the larger size having a core of unburnt material, or both. In addition, uneven firing will result, because the draught in its passage up the shaft will select the easiest path forming chimneys (channelling) and firing the material around the chimneys more than elsewhere.

Strength of stone: – The vertical load of a tall column of material will crush a soft stone or, if not crushed, a tall column of soft stone may create an excess amount of dust prohibiting successful firing, due to abrasion in its passage down the shaft. Special care must be taken to avoid these effects by reducing the length of the kiln shaft as is necessary. Firing of marl or chalk, both of which are soft stones, will require short kiln shafts.

3.3.1.2 Density, porosity and crystal structure

Effect on calcination temperature and firing: Limestone including magnesian limestones may be highly or poorly crystalline, i.e. have a well defined crystal structure or not. It may be coarse or fine grained, dense or porous. Limestone is a sedimentary rock which, when metamorphosed becomes highly crystalline, e.g. marble or dolomite.

A highly crystalline limestone is hard, fine or coarse grained material, of which the coarse grained varieties are more dense than the fine. A dense, coarse grained stone requires a higher calcination temperature and/or a longer firing time than the fine grained. Poorly crystalline rock is generally softer e.g. calcrete, and comparatively more porous than the highly crystalline stone. Porous stone requires a lower calcining temperature and/or shorter firing time than do the more dense varieties.

Shrinkage – overburning: – The porosity of the stone is reduced with a temperature increase above the calcination temperature or an extended firing time. The stone shrinks (becomes overburnt) by the increasing of crystal size and hence the reduction in the number and size of pores. The typical characteristics of an overburnt stone are low reactivity requiring long and careful hydration, and low plasticity.

Dense, coarse grained crystalline stone becomes overburnt more easily than the fine grained or porous stone and therefore requires more careful firing.

The effect of impurities: – Impurities cause slags to form around the surface of the stone which result in a reduction in porosity (overburning) at lower temperatures than would otherwise be the case. Impurities both in the stone and in the fuel have this effect but the purity of the stone is more important.

Decrepitation or spalling: – It has been found that both crystal size and structure have an effect on whether a stone will disintegrate (decrepitate or spall) during firing. In general the coarse crystal variety is more susceptible to fracturing and disintegration. However, the research results are not conclusive. Therefore the only safe way of determining whether decrepitation occurs or not is by testing the stone, preferably under field conditions. The chemical composition of the stone does not affect its structural stability during firing.

3.3.1.3 Heat and temperature

Heat is added to limestone to remove the CO_2 (carbon dioxide) and form quicklime.

The rate at which the stone is heated to bring it to the temperature at which CO_2 begins to dissociate itself from the limestone (dissociation temperature), the temperature at which the stone is fired (calcining temperature) and the length of time the stone lump is kept at the calcining temperature (firing time), all contribute to the efficiency of the firing process both from productivity and quality points of view.

Temperature of dissociation: – The temperatures at which CO_2 dissociates from limestone depends on the proportionate amounts of $MgCO_3$ and $CaCO_3$, and on the crystallinity of the stone. The $MgCO_3$ begins to decompose at a lower temperature than $CaCO_3$, a good average being around 725 °C and 900 °C respectively. Dense, coarse crystalline limestones have a higher dissociation temperature than porous, fine grained or poorly crystalline types.

Once the dissociation temperature is reached CO_2 begins to be released at the surface of the stone lump. If the temperature is maintained dissociation proceeds uniformly inward towards the centre of the lump. In practice, for the CO_2 at the core of the stone lump to be released it requires a raising of the temperature above the minimum dissociation temperature. The larger the stone size, the greater the increase in temperature required above the dissociation temperature. Boynton suggests that "the difference between the dissociation temperature of the surface and core may be 300–700 F (150–370 °C), depending primarily on the stone's diameter." (p. 161) Further factors influencing the calcination temperatures are the CO_2 pressure and concentration and the firing time.

CO_2 pressure and concentration: – A dense stone will require a higher temperature of calcination because the CO_2 will encounter greater difficulty in reaching the surface of the lump, i.e. a greater CO_2 pressure is necessary than would be the case if the stone is porous. Similiarly, a large stone, as compared to a small one of the same material, will

require a higher CO_2 pressure, i.e. a higher temperature, to calcinate. Further, if the atmosphere in the kiln is such that the % CO_2 concentration is low, the dissociation temperature is reduced.

For practical purposes, the important thing to recognize is that an interrelationship between CO_2 pressure, % CO_2 concentration and dissociation temperature does exist, and that once the optimum temperature is determined (by trial and error or by design), the firing conditions must be kept the same to ensure a consistent quality.

Calcination temperature and firing time: – At any one particular temperature above the minimum necessary, the stone must remain at that temperature for a length of time sufficiently long to dissociate all the CO_2 from the stone. The higher the temperature above the minimum that the stone is fired at, the shorter the firing time necessary. However, depending on the stone size and type, at a certain temperature level the stone will begin to overburn to an unacceptable extent. Ideally, the firing temperature should be set so as to minimize the firing time, i.e. maximize output, without reducing the quality of the lime produced, i.e. overburning it. Boynton suggests that 1149 °C for high calcium limes and 1066 °C for dolomitic limes will probably be suitable calcining temperatures. The optimum temperature can best be determined by experimentation.

Rate of heating: – Ideally, the limestone should be sufficiently preheated so as to bring the stone gradually to the dissociation temperature, and then the calcining temperature, at which it should be kept for a minimum time period.

The rapid increase of the stone's temperature to its calcining temperature is considered to have an adverse effect on quality. Practically, this means that the kiln must be designed in such a way so as to maximize preheating, i.e. the shaft should be as long as possible.

Calcination of dolomite: – Dolomite is particularly difficult to fire satisfactorily since the temperatures necessary to dissociate the CO_2 from the $MgCO_3$ component of dolomite are lower than those necessary for the $CaCO_3$ component. Firing at the temperature required for $CaCO_3$ will, to a greater or lesser extent, overburn the $MgCO_3$ component making it difficult to hydrate. It is recommended that if a dolomite has to be used it should be fired at as low a temperature as possible. Using a small mixed feed kiln of the KVI-type a temperature of around 950 °C will probably suffice for a small stone size. If a choice in fuel does exist wood is preferable since on being fired it will produce steam which helps to cool the fire and hence avoid overburning to some extent.

3.3.2 Kiln designs

Kiln designs and methods of operation (technology) differ in:

a) the manner in which fuel is used to bring the stone to the required temperature, and

b) the conditions which the stone is subjected to during the course of being fired.

The technology used depends on the:

a) physical and chemical characteristics of the stone used,

b) fuels used,

c) socio-economic and technical factors in the particular circumstances.

The objective of the designer should be to design a limeburning operation which will provide:

a) the qualities of lime required,

b) a lime of sufficiently low cost,

c) an operation which suits the local conditions.

Although it is difficult for a layman to design and determine the optimum operating conditions of a kiln, under the circumstances where a small batch or continuous mixed feed operation is appropriate and feasible, the layman could implement it with a general knowledge of the subject and a minimum of specialist assistance.

3.3.2.1 Batch fired kilns

These are kilns in which one volume of stone is fired at a time. The kiln is fired, allowed to cool, the quicklime is extracted, stone is reloaded and the kiln fired again.
Appart from the open air method of firing stone in layers with wood, batch kilns are the oldest and most well tried method. They are still successfully used in many parts of the world.

There are two types of batch fired kilns:
– the flare or updraught kiln where the stone and fuel are kept separate, and
– the "mixed-feed" type where the stone and fuel are loaded in alternate layers in the kiln.

Updraught kilns are either heavy stone structures or are built into the side of a hill, traditionally conical in shape, with a height of around two times the maximum width. The opening at the top for loading is around 1/3 of the internal diameter, and at the base, the opening for extraction is about 1/4 of

Honduran Type Kiln (after sketch in Bessey)

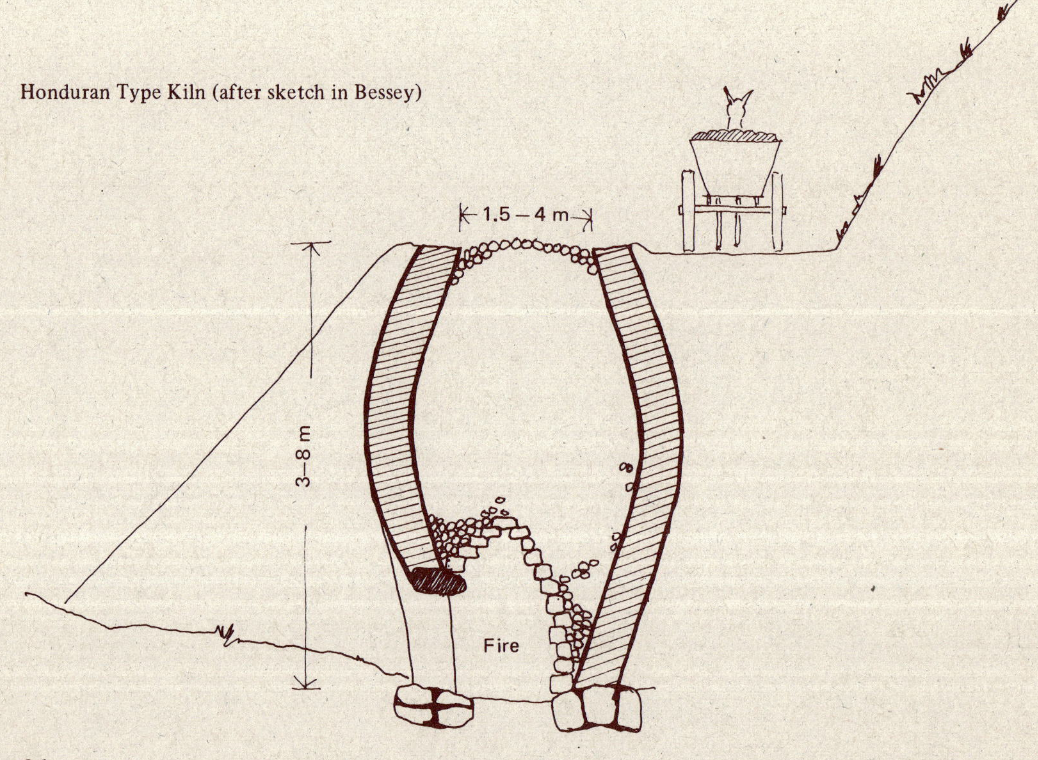

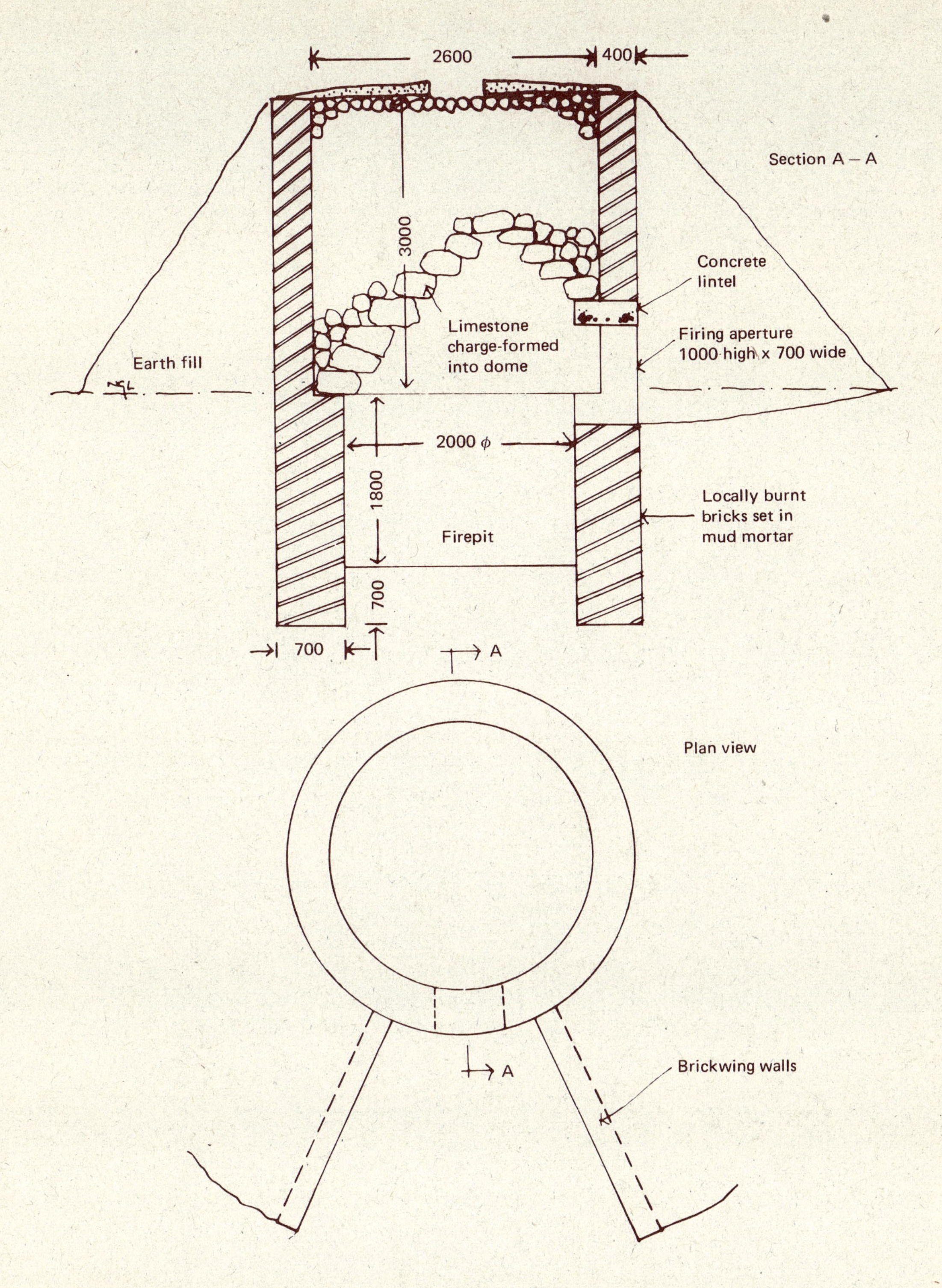

Example of Kilns Used on the Medani-Sennar-Kosti Road in Sudan

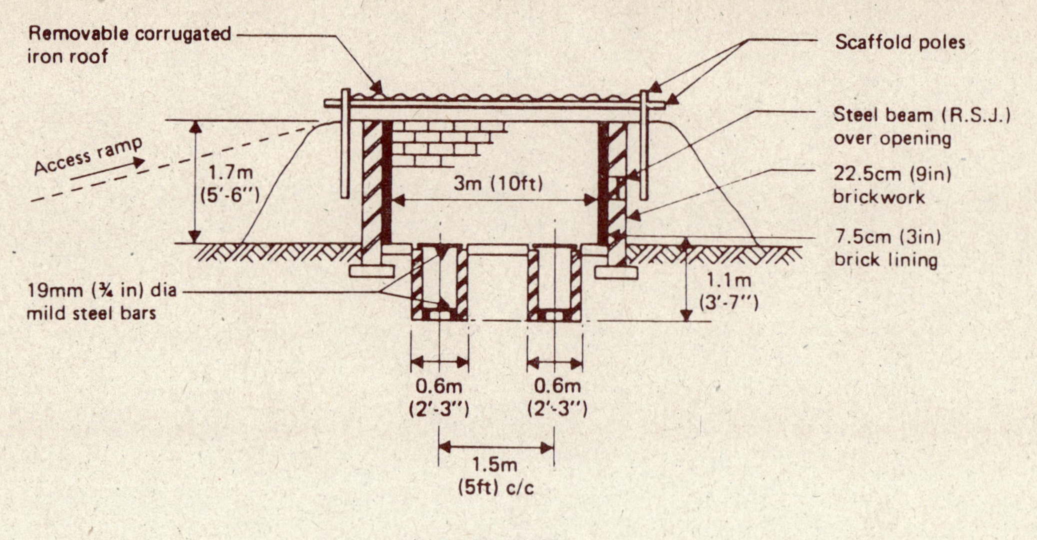

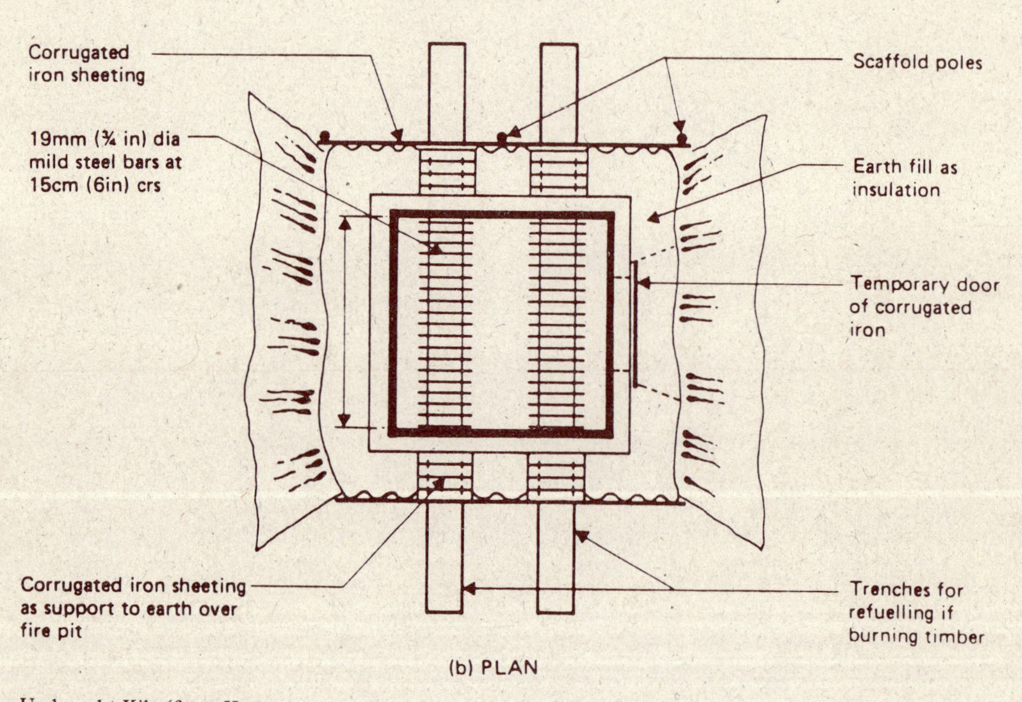

Updraught Kiln (from Hodges)

the diameter. A typical kiln of this type is the Honduran kiln. The modern versions of updraught kilns are similar to the kilns used in the Medani-Sennar-Kosti-Road-Project in Sudan, or the box type updraught kiln such as the one developed by Hodges.

Firing is carried out in a firing chamber which is either re-built with each batch or which is a permanent feature of the structure. The former is an arch or dome built of large pieces of limestone over a grate into which the ash from the fire can fall. Limestone of a progressively smaller size is loaded above the firing chamber to the top of the kiln. The method of loading and the size of stone used, which depends on the fuel used and the kiln design, are very important factors for the production of good quality lime.

Where the firing chambers are a permanent feature, such as in the Hodges updraught kiln, special trenches are built into the kiln. The way in which stones of different sizes are arranged in the kiln is equally important with this type of firing chamber. In both cases firing is very much a matter of trial and error and can best be learnt by experience.

In both methods there are no means of ascertaining when the limestone has been completely burnt except indirectly by observing, through the inspection or poking holes, the colour of the limestone during firing which should be a bright cherry red when the right temperature is reached. Usually firing is continued for 48 hours after the fire has reached the maximum temperature (cherry red colour). This means about 72 hours per batch in total. A large proportion of overburnt and underburnt material can be expected (around 25 % of each).

Mixed feed type batch kilns are most commonly used in the rural areas of developing countries, particularly because of their mode of operation. They require relatively little attention during firing. In design they are similar to those used for continuous mixed feed operations. They vary mainly in the method of operation. Limestone and fuel are fed in alternate layers up the kiln shaft, in the required proportions, fired and allowed to cool. Then the product is extracted.

Compared with kilns which are operated continuously they are wasteful of heat. This is a very important factor for consideration particularly in a situation where fuel is scarce, expensive or a long distance from the production site. The economy of fuel is low in such kilns because of the huge waste of heat in raising the temperature of the kiln walls each time a new batch is started, and also because of the loss of heat to the atmosphere when the fire reaches the upper limits of the kiln.

The attractive feature of such a kiln for developing countries is that it requires very little attention during firing, making it suitable in situations where labour is scarce or its availability restricted, or where demand is irregular.

For example, its use would be advantageous in a distant rural road project where lime is required for soil stabilization and which is a long distance from a lime producing centre. Under such circumstances the erection and operation of a small lime kiln (or kilns) near the road construction site might be preferable. The method of ascertaining whether the material is completely burnt and the length of time for each batch is the same as in updraught kilns. Kiln designs of this type are like those of the 'Somali kiln', the kiln used in Papua – New Guinea and the 'Khadi Village Industries' type kiln. (See section 3.3.3)

The principles and details of kiln construction and materials suitable for use which apply to the continuous type kilns apply to the batch kilns as well.

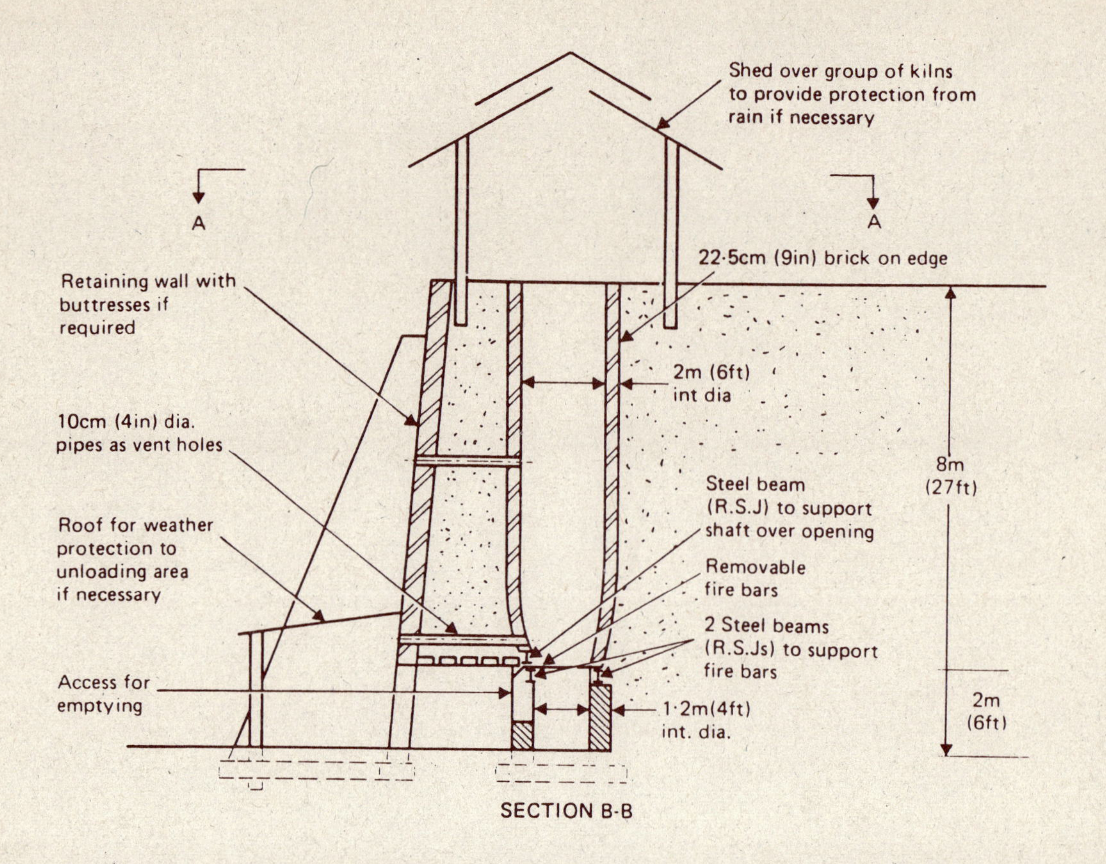

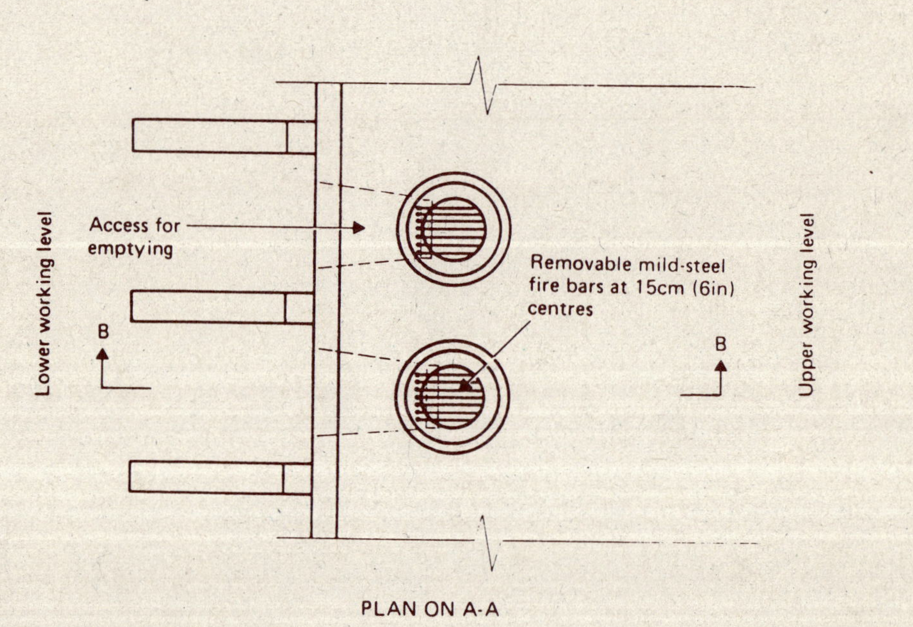

Somali Kiln (from Bessey)

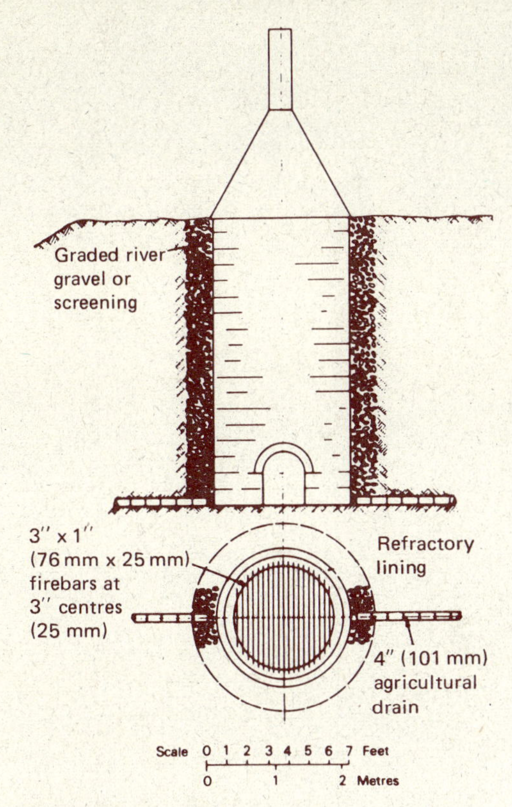

A Simple Shaft Kiln for Wood-Firing
(after Hoskings)

3.3.2.2 Continuous kilns

There are two different types of kilns which may be suitable for a low demand situation; the vertical shaft kiln and the horizontal kiln.

Vertical shaft kilns are symmetrical, either circular, elliptical or square in shape and are fired either internally or externally. Mixed feed kilns are the internally fired type and are recommended for use wherever a very high quality product is not required. The externally fired types are more difficult to construct, and for efficient and effective use require special design and operating effort. The installation of such a kiln to fire a limestone of less than 95 % $CaCO_3$ is considered wasted effort since the quality lime produced will not be sufficiently high to sell as high grade lime which would warrant the higher capital and fuel cost. Externally fired kilns, particularly if using low technology methods, generally consume more fuel than the mixed feed type.

In mixed feed continuously operated kilns (internally fired) solid fuel such as wood, coke and coal is charged in alternate layers with limestone at the top of the kiln. The fuel is burnt in the middle third of the kiln shaft to calcinate the limestone. The rising hot gases from the combustion in the firing zone (middle third) preheat the limestone in the top 1/3 of the shaft whilst the material in the lower third is cooled so that it may be comfortably withdrawn through discharge openings at the base of the kiln shaft. For the production of lime suitable for construction purposes, soil stabilization or agricultural use, the mixed feed type kilns are likely to be the most appropriate. They are therefore dealt with in detail in sections 3.3.3–3.3.4. (See also the drawing of the construction of a vertical shaft kiln p. 54. This is a KVI-type kiln used in Moshaneng, Botswana with recommended variations). A kiln design used in India, as described by R. Spence, has been included below (p. 42).

Solid fuel, furnace kilns are modernized versions of the traditional updraught kiln. Wood, coke or coal can be burnt in fireplaces built into the wall of the kiln approximately 1/3 the distance up the shaft. It is a method which is most wasteful of fuel and very difficult to operate to produce a consistently good quality quicklime. There is a tendency to fire the limestone at the periphery of the shaft more than at the centre. To limit this, the shaft diameter must be kept to a minimum and if possible, either by the injection of steam or re-cycled flue gases into the

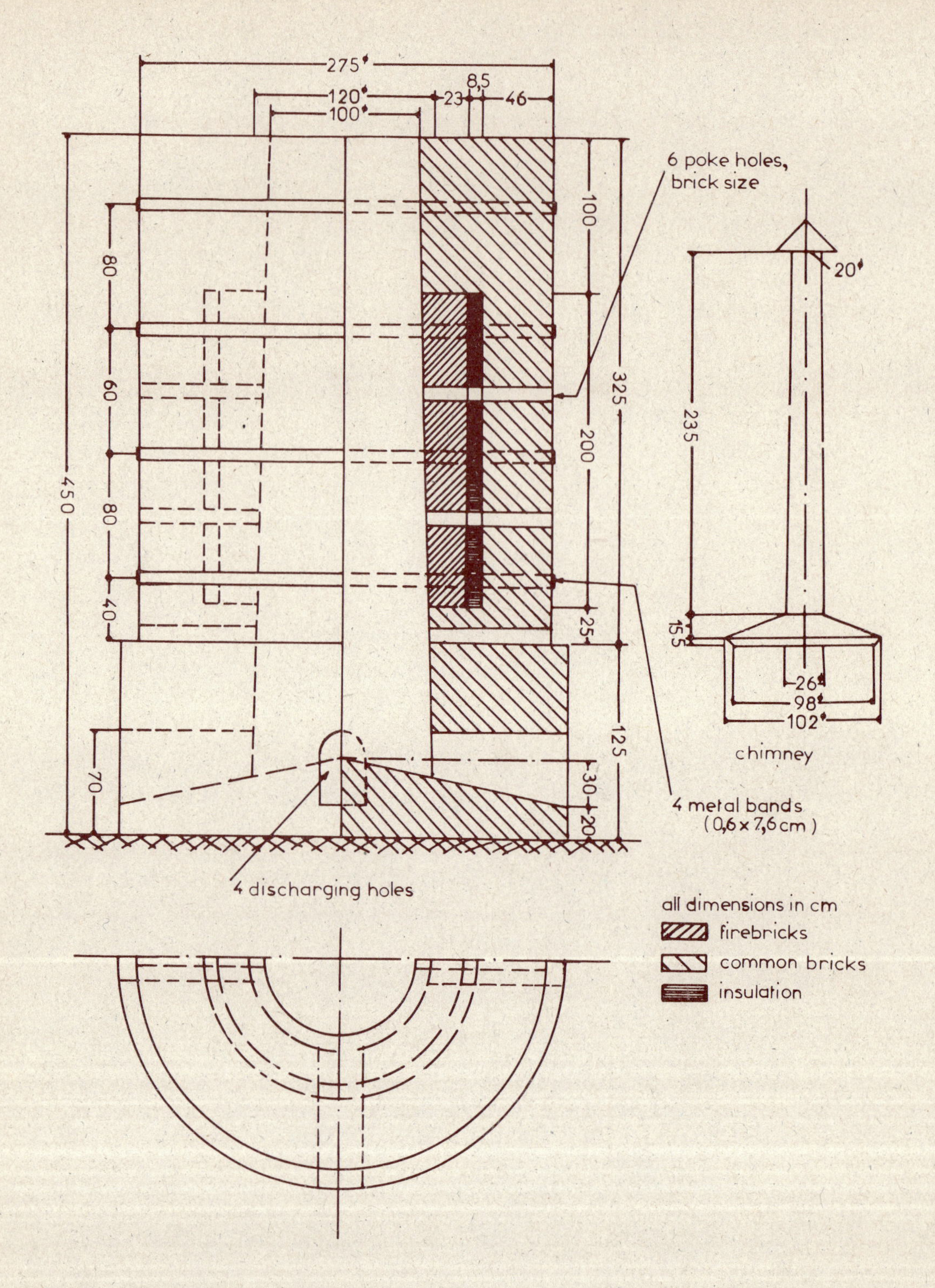

Kiln for Quicklime Burning (from Spence)

fireplace, the flame should be forced towards the centre of the column of limestone. A kiln fired in such a way produces the best results when wood is used as fuel since this has the advantage of producing the long, cool flame which is considered to be most suitable for limeburning. The risk of overburning is avoided. If coal or coke, which produce shorter and more intensive flames are used, the injection of steam or recycled flue gases will lengthen and cool the flame. Volatile coal, which is not suitable for mixing with limestone, could be used if special care is taken. The difference in kiln design when coal is used as the fuel instead of wood, is the size of the furnace. Since wood is a bulkier material it will require a larger furnace.

Sketch of Fireplace/Furnace for Externally Fired Kiln

Dimensions have not been given since the fireplace will have to be specially designed in each case, depending on the fuel used or the limestone fired. The fireplace must be lined with firebricks and should be insulated. The water container shown in the drawing may be omitted. It is a means of utilizing the heat loss to the furnace walls but its construction and maintenance may be too difficult and costly.

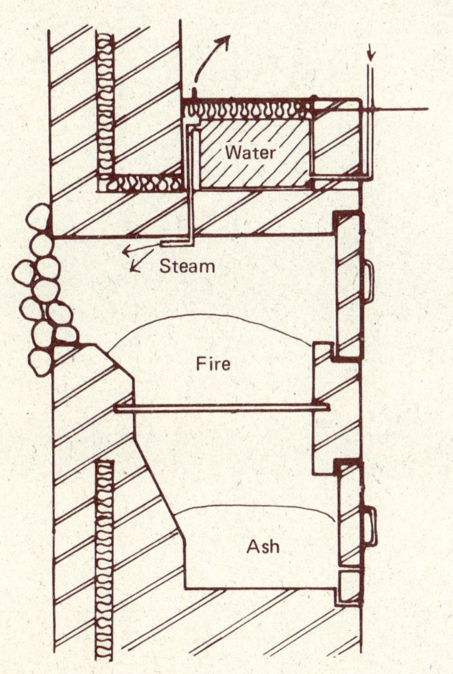

Producer gas fired kilns: – The fuel efficiency of furnace kilns can be improved if the fuel is converted to producer gas before being fired. The fuel saving factor is that the combustion of the fuel takes place inside the kiln shaft as opposed to in the fireplace. Gas can be produced in gas producers either built into the kiln wall or as separate structures. Although more difficult to construct and also to operate, gas producers built into the kiln wall will guarantee far less heat loss than producers separated from the kiln. It is difficult to say whether the one is preferable to the other. The implications should be considered in each instance. The design of the kilns having gas producers built into the kiln structure is similar to the externally fired kilns with the difference that the bed of fuel in the fireplace is deeper and provision is made to allow steam and air to pass through it. The diameter of the kiln

Gas from the producer rises vertically up the gas duct into the mixing duct as does air up the air duct. The two are mixed under a slight pressure before being ignited. The distance of the two ducts from each other and the length of the mixing duct determine the nature of the flame. Either primary or secondary air can be used. The larger the mixing duct the more intensive the flame. (Sketch from Searle, p. 313)

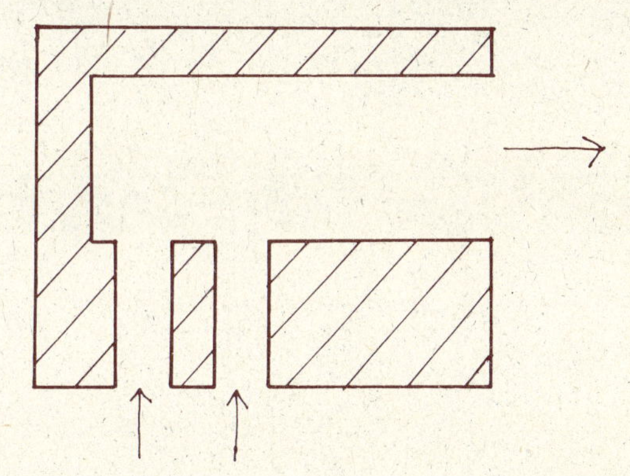

shaft should be a maximum of two meters. If a greater diameter is used it will be difficult to fire evenly right across the shaft cross section. Whether one single producer or several small ones are used, the gas should be introduced at about 1/3 of the height of the kiln and through as many inlets as possible. Air is mixed with the gas either in the producer (primary air), or in the kiln with air which has passed through the cooling zone of the kiln (secondary air), or with both. The producer gas is ignited in the firing zone by the hot secondary air or, if only primary air is used, it is introduced under pressure with the gas into a mixing pipe the end of which is ignited. The proportioning of the two types of air and gas and the pressure at which the mixture is introduced as well as the burner design determine the quality of firing (see diagram of burner for gas firing described by Searle). A well designed and operated gas fired kiln can produce the best quality lime. The gas producer enables the use of fuels which would otherwise be of little value. Lignite, peat and high volatility coal as well as certain agricultural waste material, such as coconut, coffee husk, olive pips etc. can be used to produce gas. Good design and operation of the producer is vitally important. As a general principle the simpler the producer the less likely it is to break down or operate inconsistently.

Oil fired kilns: – In the simplest form oil is burnt in combustion chambers in the wall of the kiln at around the bottom of the middle third of the shaft. The kiln is easier to construct and operate than the gas fired kiln and will produce approximately the same quality product. Heavy oil or even used motor oil can be used. It is atomized, mixed with the right proportions of air or steam and ignited.

Accurate firing requires the regulation of the mixture of air and oil or steam and oil to ensure a suitable length and intensity of flame. An intensive flame will tend to overburn the limestone whereas a very long flame with an inadequate amount of air will tend to choke the pores of the limestone with soot and will not reach the centre of the column of material.

Injection System Used in Aspropirgo, Greece

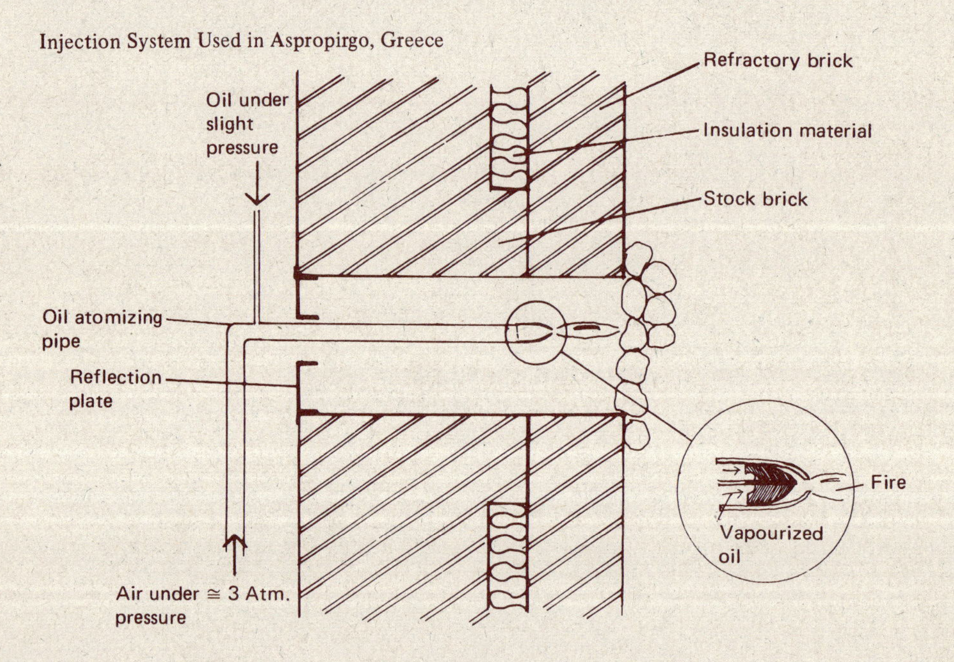

Indonesian Oil Fired Injection Kilns
(after Sobek)

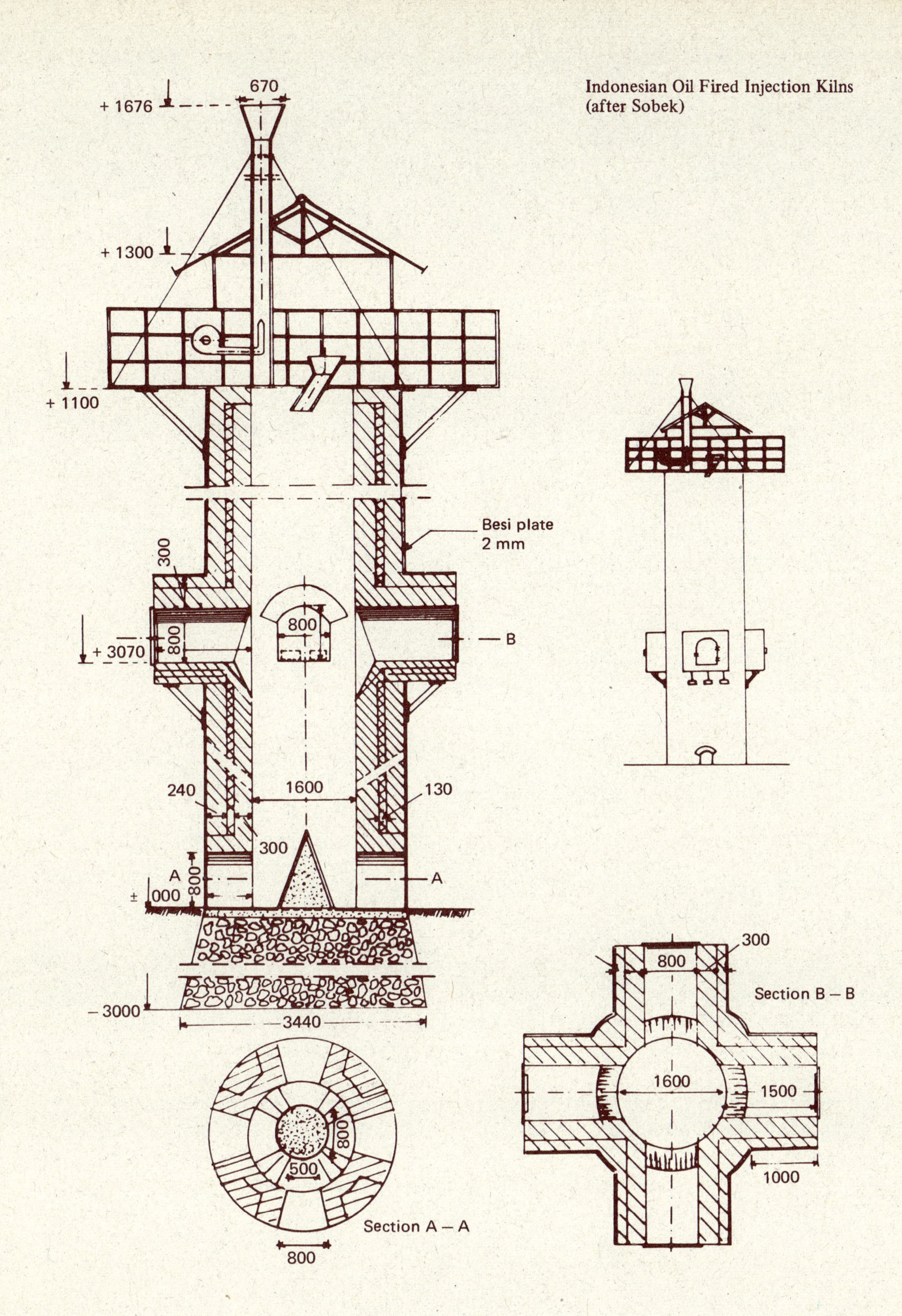

Greek Oil Fired Kiln

Horizontal kilns: – There are three different types of horizontal kilns: ring kilns, rotary kilns and tunnel kilns. Ring kilns are of specific interest for situations requiring a low technology solution.

A ring kiln or Hoffman kiln can be either circular or elliptical in shape. It is a continuous tunnel structure into which limestone is charged and from which quicklime is extracted continuously. Hot gases are drawn horizontally through the limestone to produce the quicklime. To describe its operation, let us assume that the kiln is divided into 20 chambers.
Under ideal operating conditions:
1 chamber will be empty,
1 chamber will be filling,
7 chambers will be pre-heating,
4 chambers will be under fire,
6 chambers will be cooling,
1 chamber will be emptying.
The larger the chambers the more efficient the firing process and the efficiency of the kiln. The limestone is carefully stacked by hand with hollow vertical shafts formed directly below the fuel feeding apertures in the roof of the kiln to take the fuel. Coal, oil or wood can be used as a fuel which is ignited by the very hot gases flowing horizontally through the kiln. The fire is drawn forward by opening and closing draught-inducing caps in the successive chambers. The air is drawn through openings in the external wall of the kiln to a 15 m chimney.
The permanent roof of the kiln may be omitted and substituted by a double layer of bricks covered with sand or ash. This will save on capital cost but makes for a less fuel efficient kiln and one where the lime is susceptible to damage by rain. This roofless kiln can be constructed below ground level and has the advantage that the limestone can be loaded and extracted using a mechanical grab.
The advantages of using ring kilns are that they have a low fuel consumption and can produce a cleaner lime than that produced in a mixed feed vertical kiln. The disadvantages are that the labour cost is high, they require careful attention in operation and have a high capital cost.

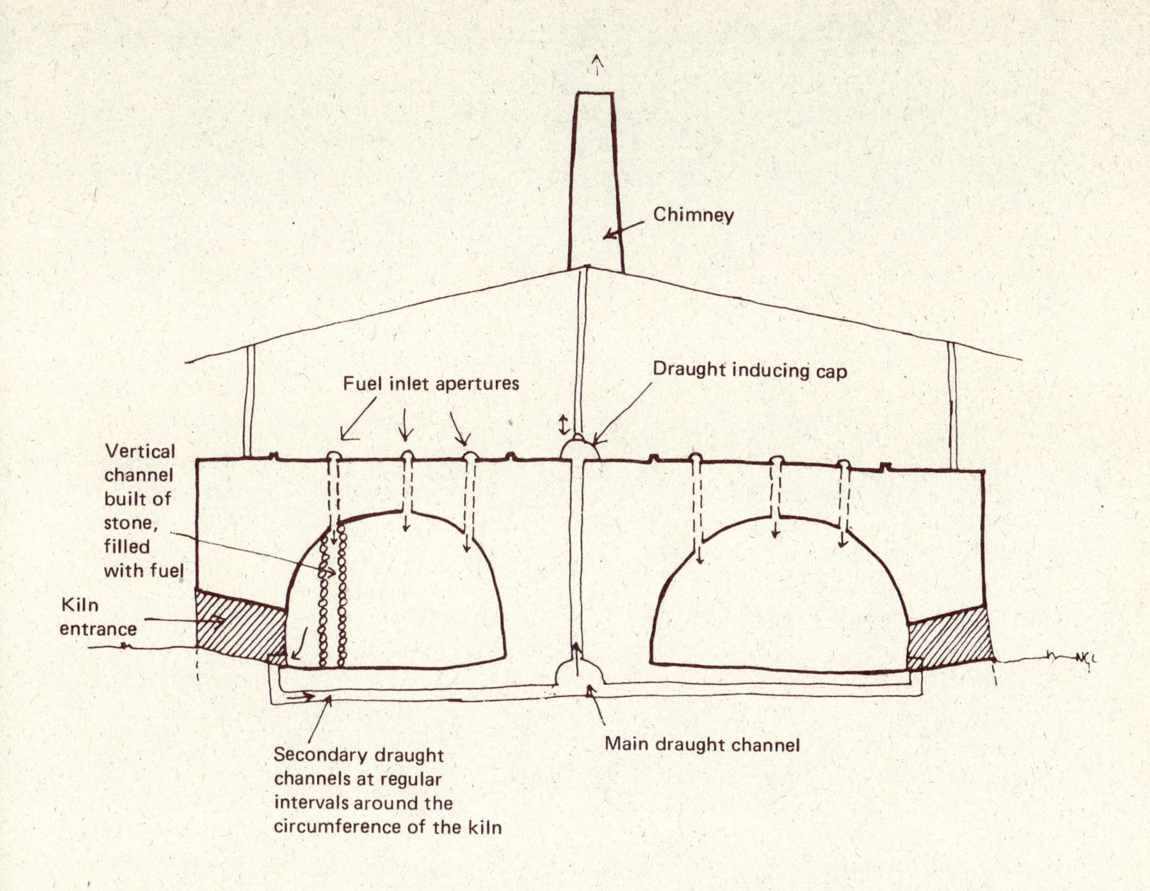

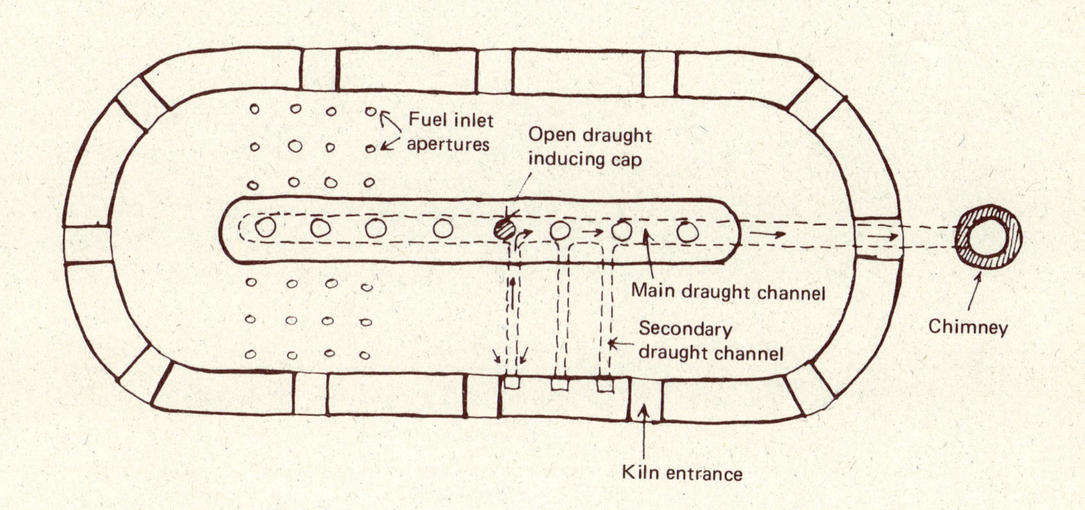

Hoffman Kiln, Observed in Greece

3.3.3 Vertical shaft mixed feed kilns, design and construction

Of all the various shapes that have been used, the simple cylindrical shaft type is the easiest to construct and operate. It is likely to be suitable for most situations. Alternate layers of stone and fuel are fed at the top, fired around the middle third and quicklime is withdrawn from openings at the base.
Searle describes the ideal mixed feed kiln as, "... sufficiently high to make the fullest possible use of heat from the fuel and to cool the lime to such an extent that it can be handled with ease. It must be such a shape that the stone and lime will pass easily down through the kiln without any poking or other attention being necessary, and, at the same time, the lime must be burnt uniformly, none of it being spoiled by overheating and none being incompletely burned.
The manner in which the stone and fuel are supplied to the kiln should require the minimum labour, and the withdrawal of the burnt lime should be effected in the simplest possible manner," (p. 281)

3.3.3.1 Shape and dimensions

The circular in plan, perfectly vertical wall shaft is easy to construct, makes for a better distribution of heat than the square or oval plan types, and also reduces the potential for the material to "hang" in the shaft. The oval plan shafts are best used in externally fired kilns and the square types, although easy to construct, are likely to have problems of heat distribution to the corners. A possible departure from the simple cylindrical shaft kiln is one which tapers slightly to the apex. This shape will reduce the amount of hanging.

The kiln should be as high as possible to maximize both its preheating and cooling zones. This will reduce the heat losses to the atmosphere in the hot exhaust gases and the hot lumps withdrawn. Besides the fuel saving implications, the length of the kiln affects the draught through it.
The kiln height is limited by the draught required, the stone feed strength and the stone's resistance to abrasion during its passage down the shaft. A soft stone such as chalk could not be fired in a tall kiln since the individual lumps would be crushed under the weight of the column of material. Also, a stone type which does not resist abrasion will create excess dust in the kiln inhibiting the fire. The taller the shaft, the greater the abrasion and dust created. Searle suggests that whereas the height of the kiln depends on the physical characteristics of the stone feed, the diameter depends on the output required. However, as a general rule a ratio diameter to height of 1: 4 can be recommended for a porous stone feed and 1: 4.5 for a dense type.

3.3.3.2 Kiln construction

The kiln must be designed and constructed so that not only will the highest possible quality lime at the lowest cost be attained but also that:

a) it is a strong, stable structure which will require a minimum of maintenance and repairs;
b) it has an easy and safe access for repairs and maintenance work and allows for safe and comfortable operation.
The most economically suitable materials should be used, bearing in mind the social and environmental implications of the choice.

The *kiln base* must be constructed on firm ground and of a size adequate to take the load of the shaft above and of the kiln contents. It should have as safe and as comfortable a working area as is possible for the discharging of the kiln. It can be made of concrete, masonry or brickwork.

The functions of *discharge openings* are firstly to allow quicklime out of the kiln and secondly to allow air into it. There are two types of discharge openings. In the one the burnt stone flows out from the centre of the column of material, and in the other the material flows inward towards the centre of the column.

The inflow type has the following advantages over the outflow:

a) It makes for a more even draught because it is not affected by the direction of the wind. In an outflow type the draught is strongest on the windward side of the column of material. If the wind is strong enough and blows consistently on the one side it will cause uneven burning.

b) Draught control is easiest with an inflow type since a portion of the opening can easily be blocked off to the necessary degree (as determined after testing), without inhibiting the extraction of quicklime.

c) Extraction of material will require less manual effort since the stone will not need to be shovelled out.

The main disadvantage is that construction is more complicated and probably more expensive. Further, if the opening size is not sufficiently large it will cause the stone lumps to become stuck requiring prodding to release them to continue the discharging. An opening 1/3 of the diameter will probably be sufficient.

An outward flow type opening will be easier and cheaper to construct, the flow of stone out of the kiln will be easier and more even but will require more manual effort. The main problem with this type is the control of draught.

The inflow type is particularly useful if a kiln is built into a hillside or embankment whereas the outflow can best be used in a free standing kiln.

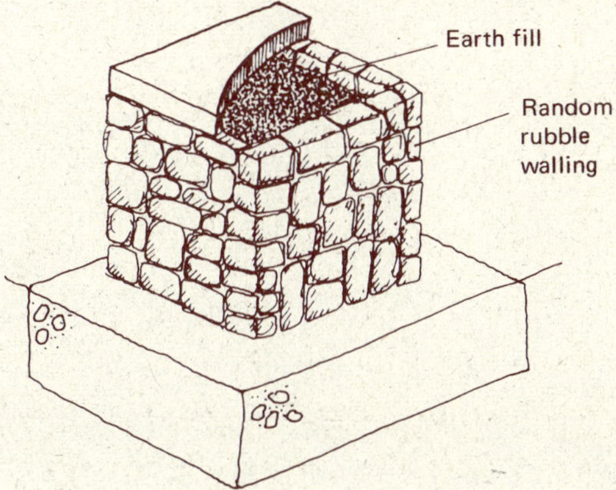

Base for outflow type discharge opening

Base for inflow type discharge opening

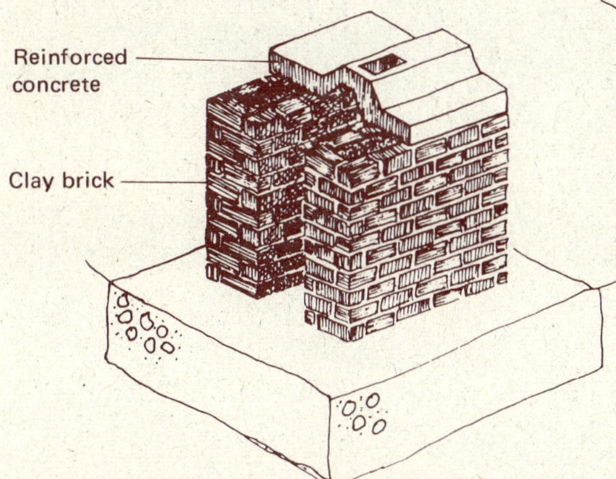

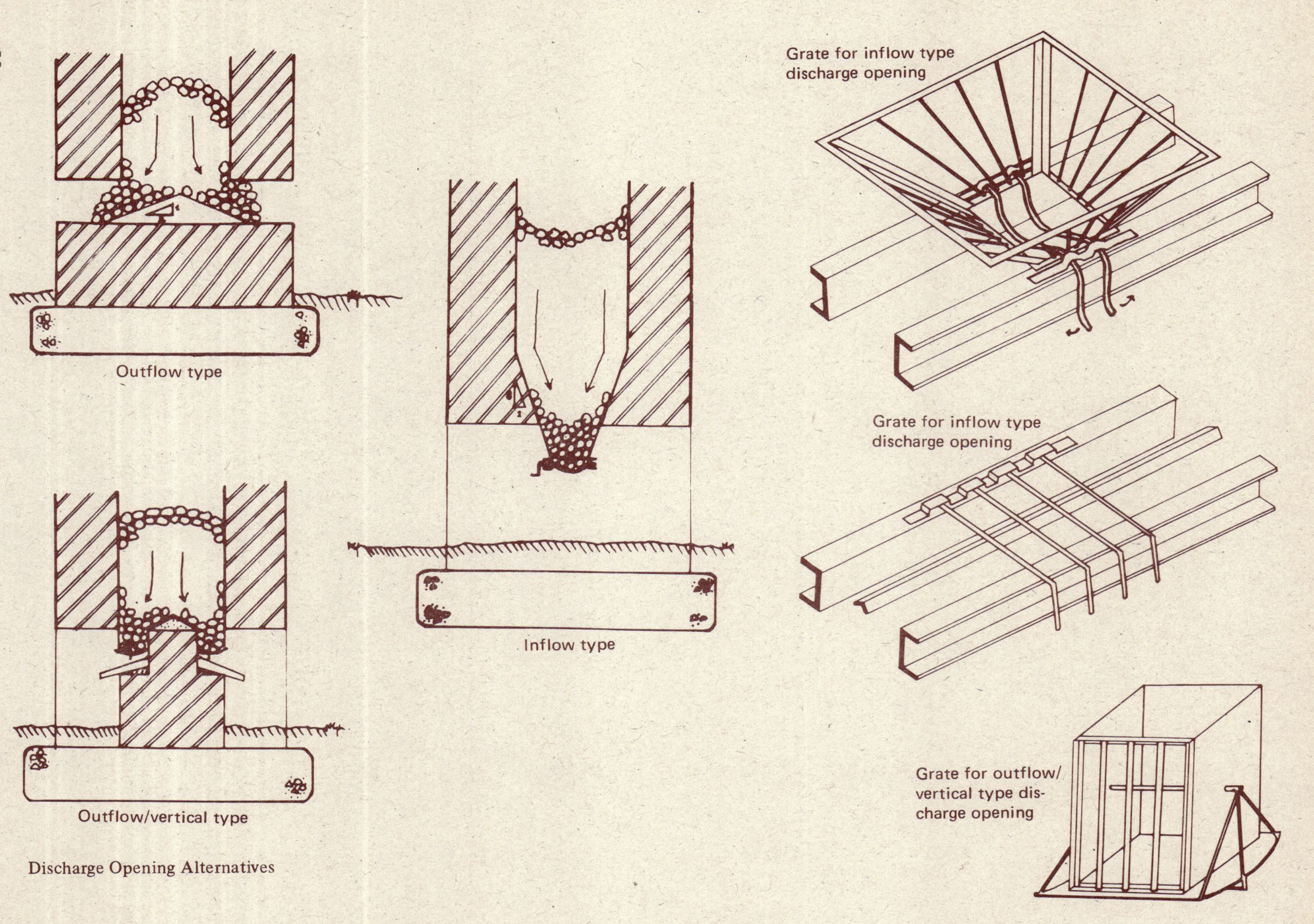

Discharge Opening Alternatives

In addition to supporting its own weight and all sundry structures, the *kiln wall* must support the lateral pressure of the column of material and must resist cracking due to heat expansion. Further, the wall construction must be such that it limits the heat loss due to radiation, and protects the inner lining from sudden changes in temperature. The walls must be sufficiently thick and made of the necessary materials to meet these ends. They are normally of masonry, concrete, brickwork or a combination of these and between 600–1000 mm in thickness depending on the materials used and the size of the kiln.

The wall should be strapped with metal bands, reinforced with mild steel bars, or thickened or buttressed to take the lateral loads.

The *inner lining* of the kiln wall is subject to the abrasion of the material as it descends down the shaft. It is also subject to wear and tear due to sudden changes in temperature, for example when opening inspection eyes or pokeholes, and due to the chemical action of the lime and the kiln gases. The upper part of the kiln must be protected against abrasion and the middle and lower against chemical action.

Hard, dense bricks are suitable for the upper quarter of the kiln, i.e. hard "blue" engineering or paving bricks, or granite worked smooth so as to reduce the wear on the loaded material in its passage down the shaft.

Hard, fine textured refractory bricks are ideal for lining the firing zone of the kiln. The high alumina type is the most satisfactory as the chemical action of the lime is reduced. A refractoriness of about 1200 °C is quite acceptable.

The lining should be carefully constructed with as small a proportion of jointing as possible. The joints between the bricks are particularly susceptible to the abrasive and chemical action of the lime, and must be constructed as thinly as possible or in no time at all bricks will start to fall out of the wall. Three quarters of the kiln shaft, from the base upward should be lined with refractory bricks.

If high alumina refractory bricks are not available hard engineering bricks may be quite suitable and if these are not available, other hard material such as sandstone or schist laid with the grain horizontal could possibly be used, or even blocks of limestone.

If these latter alternatives need to be used they should be well tried during the testing phase in the test kiln described in section 4. In any event, both the short and the long term cost implications of the alternatives must be carefully considered before a decision is made. The cost of regular replacement, bearing in mind the cost of loss of production during the replacement period, may in fact be higher than buying and transporting refactory bricks from a distance.

It is recommended that some form of *insulation* be built between the inner lining and the kiln wall (at least in the middle half of the kiln). Large amounts of heat can be lost to the walls and then to the atmosphere. Bricks made of diatomatious earth (Kieselguhr), loose diatomatious earth or some other material such as pumice stone, or even air gaps can be used as a means of containing the heat. The insulation should be built between the lining and the kiln wall forming a 50–100 mm layer between them. In addition to insulating, the material can take up movements due to expansion.

The *charging opening* does not require any special mention but what is important to note is that the working area around the opening must be sufficiently spacious to be safe and comfortable. If necessary a platform of the type drawn in the figure p. 53 should be built round the opening.

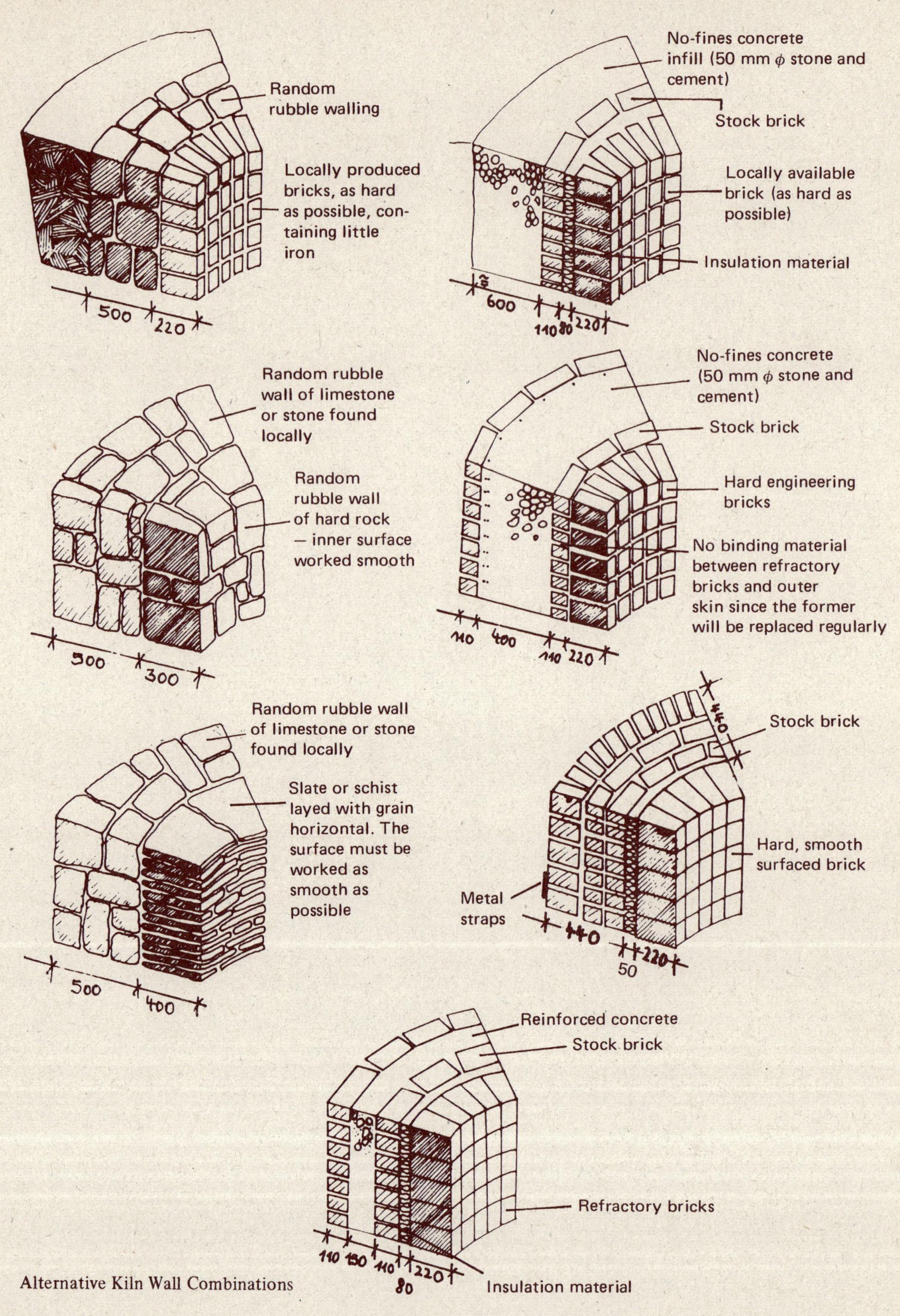

Alternative Kiln Wall Combinations

The primary function of the *chimney* is to assist the creation of the necessary draught through the kiln. Secondly the exhaust gases are drawn away from the workers feeding the kiln making their working conditions bearable.

Loading mechanisms bring the feed (fuel and stone) to the charging opening (see the fig. p. 29).

Pokeholes should be constructed into the wall of the kiln at suitable positions at the

Charging Opening Alternatives and Chimneys

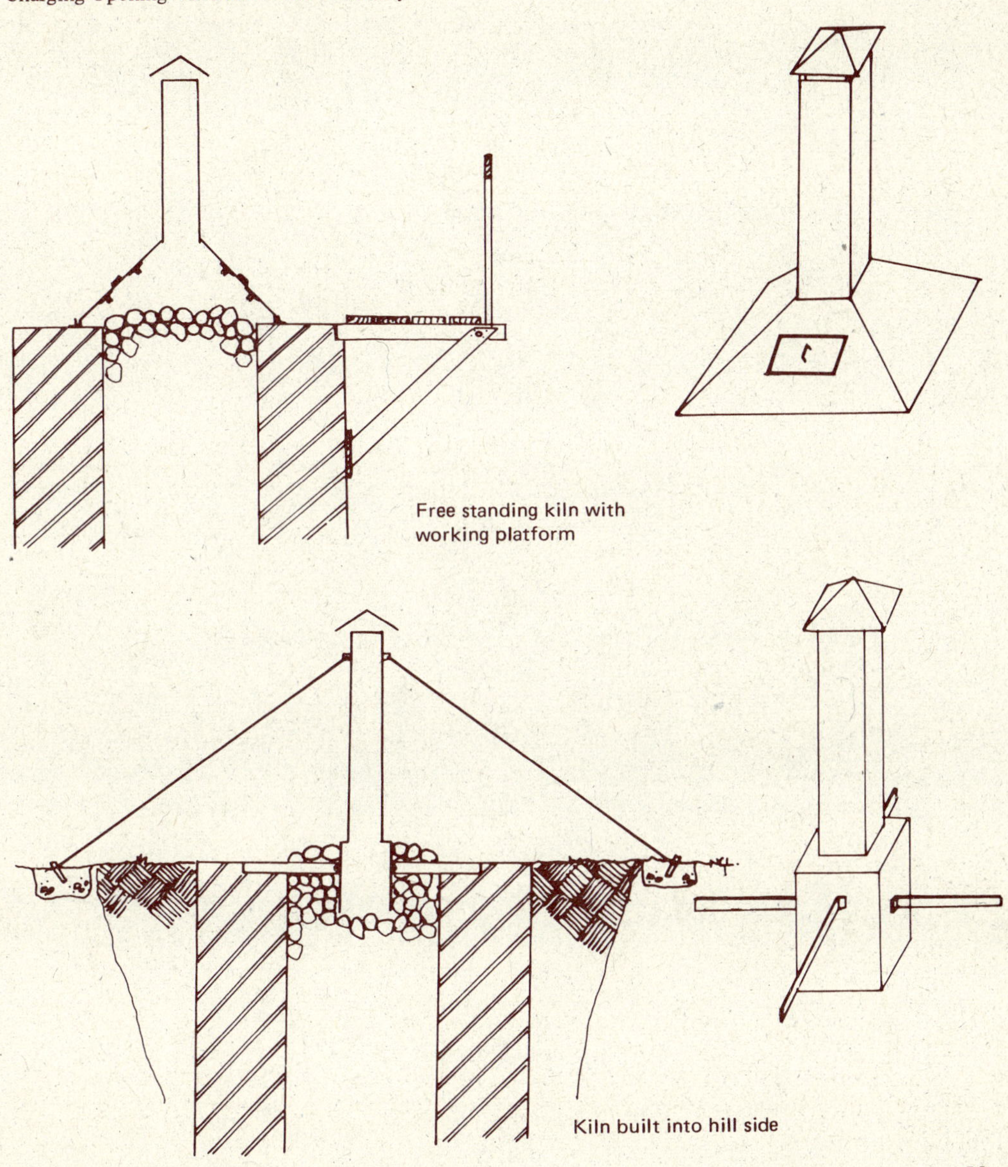

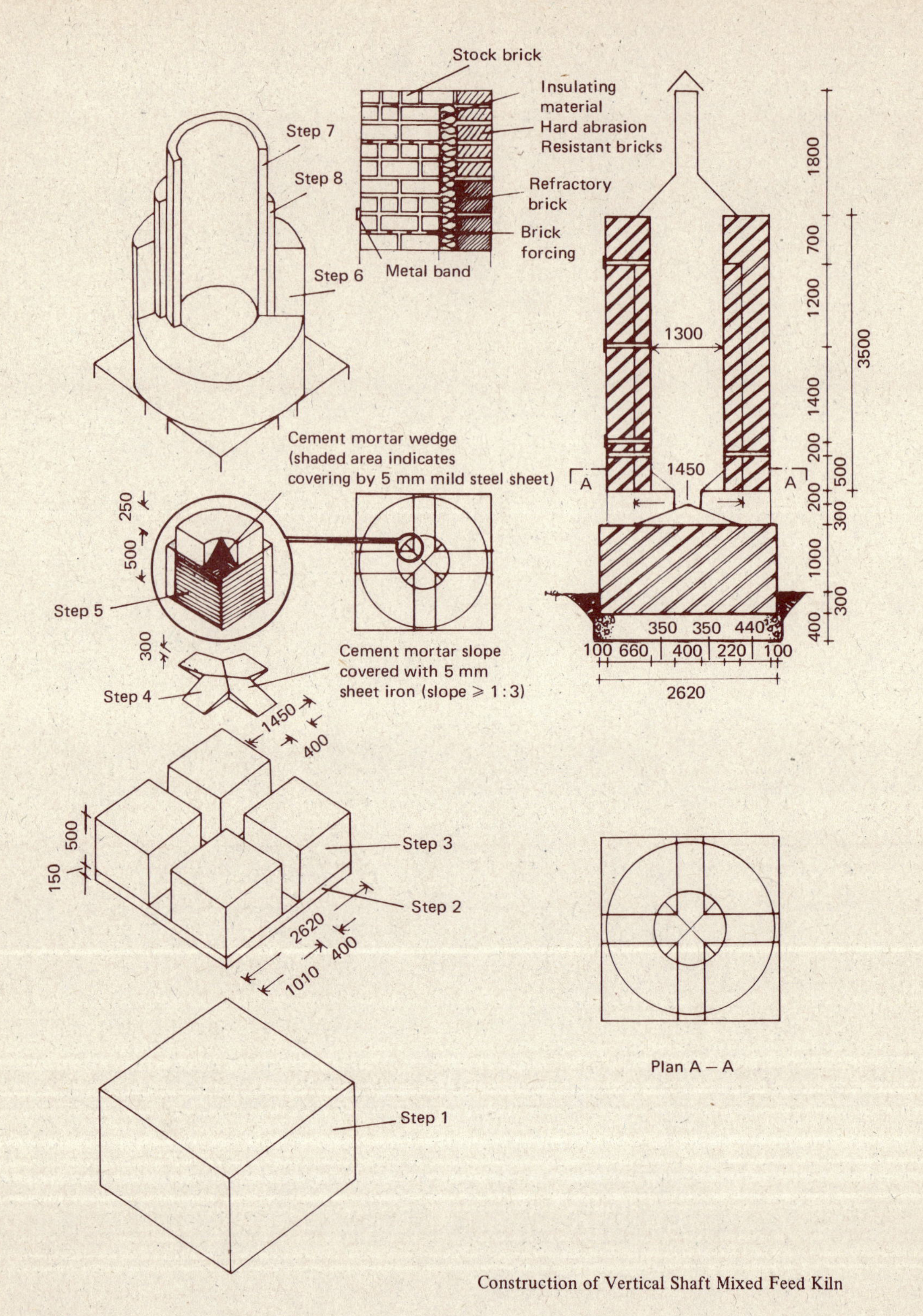

Construction of Vertical Shaft Mixed Feed Kiln

top and bottom of the firing zone to allow for poking at regular intervals to prohibit the development of "hanging" (see hanging). The pokeholes should be such that they can be easily opened and closed during firing and that they are airtight. *Inspection eyes* should be built into the wall at positions permitting the burning zone to be viewed and the temperature checked at the various levels by using a thermocouple. Other apertures that may be left in the kiln wall are airholes. These should be near the base of the cooling zone of the shaft.

3.3.4 Continuously operated, mixed feed kiln

To produce the best possible quality lime the kiln/s must be designed and operated to:
a) achieve and maintain a uniform distribution of heat,
b) achieve the necessary calcining temperature and maintain it at the required level in the kiln,
c) fire the limestone for the required time period.
To achieve these conditions, design and operation should ensure:
– an adequate draught,
– the correct loading, sizing and grading of feed,
–charging and discharging at a suitable rate.

3.3.4.1 Design factors

The purposes of the *draught* are to provide sufficient oxygen for the fire to assist the combustion of the fuel, and to cool the burnt stone (quicklime) in the lower third of the kiln so as to enable it to be handled with ease on withdrawal. The rate at which air flows through the kiln should be such that it draws the exhaust gases and the fire upward at a rate which corresponds with the length of time the stone lumps must remain in the firing zone (firing time). A short firing time will require a fast flow of air through the kiln and vice versa.
The factors that affect both the quantity and rate of air flowing through the kiln, i.e. the draught are:
– the length of the kiln,
– chimney length,
– sizing and grading of stone and location,
– size and location of openings at the base.

Length of the kiln: – Searle recommends that the kiln be as tall as possible so that both the preheating and cooling zones can be maximized. The length of the kiln however, does not serve this purpose alone but also determines the rate at which the air flows through the kiln. The taller the kiln shaft, the stronger the drawing effect upward.

Chimney length: – The draught can be increased by using a chimney above the loading opening at the top of the kiln. The extent to which it serves to increase the draught depends on its length and cross-sectional area. There are various types contributing more or less to the ease with which the kiln can be loaded, i.e. drawing exhaust gases away from operators loading the kiln.

Access Ramp and Chimney Used in Moshaneng

Sizing and grading of stone: – As stated previously the smaller the stones and the wider the range of sizes, the more difficult it is for the draught to pass through the kiln, and also the shorter the firing time necessary to achieve a well burnt quicklime. The length of the kiln and the chimney must be balanced with the size of stone used so as to get the required draught in the kiln, and hence the necessary firing time.

Charging openings/discharging ports: – The air should be drawn up from the base of the kiln to perform the cooling function, to warm the air used for combustion and to acquire the full effect of the kiln length for these purposes. There are various types of discharge outlets which vary in size and position. The size determines the amount of air let into the kiln, and the position has a qualitative effect on the air flow through the kiln and on the ease of operation.

3.3.4.2 Operational factors

Charging: – Correct charge is most important for the assurance of a uniform heat distribution in the volume of stone being fired. Stone of a consistent size and grade must be fed in the correct proportions with fuel in layers which are as fine as is practically possible. This will maximize the degree of mixing of the limestone and fuel. The practical limitation to the layer size or fineness is handling. Generally, the smaller the layers, the greater the amount of handling and the corresponding handling cost, whether manual or mechanical.
The stone feed size and grade should be checked at regular intervals, particularly if hand dressing methods are employed. The fuel feed should also be checked regularly for the same things.

Rate of charging and discharging: – The rate at which stone is charged into the kiln and withdrawn from it affects both the quality produced and the productivity. The rate at which the quicklime is extracted and limestone subsequently charged, is dependent on the rate at which the fire rises up the shaft which in turn depends on the draught and stone size selected. On average 1/3 of the limestone feed fired will be extracted from the kiln every 7–10 hours. In other words, the stone will be in the firing zone (middle third) for a period of around 7–10 hours and will be kept at its peak temperature for about 4 hours. The fire will generally rise up the shaft at between 100–150 mm per hour.

Lighting the fire: – In many instances starting a fire in a mixed feed kiln will be very difficult. Placing oil-soaked rags or wood at the base of the column of material together with some kindling and then lighting will serve to alleviate this problem.

3.3.4.3 Problems – causes and remedies

Overburning results from firing for a longer time and at a higher temperature than necessary. The result is that the stone becomes hard, it shrinks and is then slow to hydrate.

Detection:
– Overburnt stone will make a sharp ringing tone when tapped with a hammer, compared with soft burnt stone.
– A fixed volume of overburnt stone will be heavier than soft burnt stone due to shrinkage, i.e. more stone lumps will fit into the same volume.
– If water is poured over an overburnt stone and a lightly burnt stone, the former will react more slowly (low reactivity).

Remedies:
– Reduce the temperature by reducing the fuel: stone ratio (use less fuel).
– Increase the stone size.
– Reduce firing time by increasing the draught (increase chimney length).

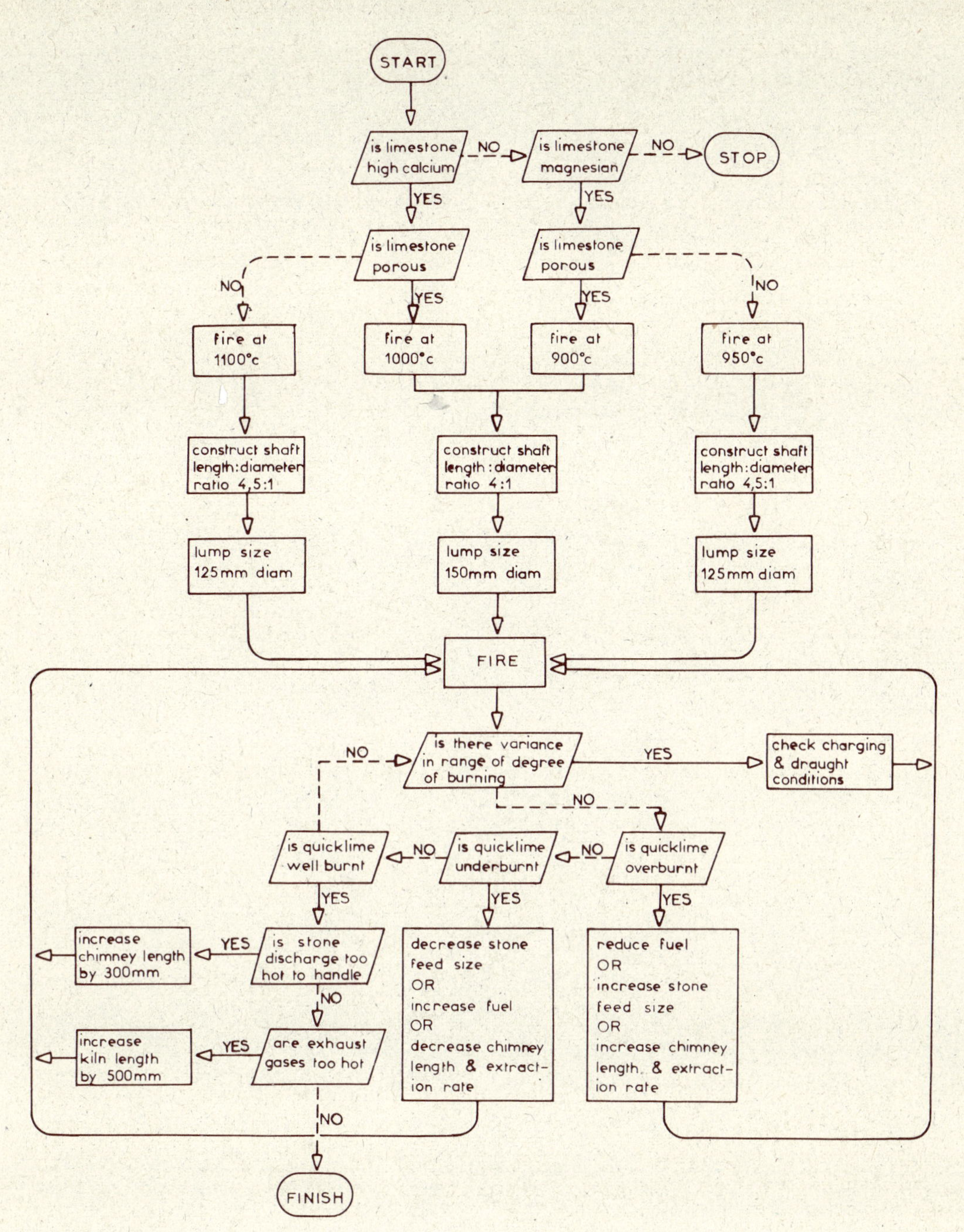

Trials Procedure

Core material results from underfiring, i.e. firing at a low temperature and for a shorter time than is necessary.

Detection:

– The individual stone lumps when broken to show the inside have a soft, well burnt outer shell which can be broken easily with an ordinary hammer blow, and a hard core which is difficult to break and possibly has a different colour.

– Another way to detect the existence of core material is by pouring water over the burnt lump. The reactive portion will fall away leaving a lump of core in the middle.

A core of 5 mm diameter can be considered as acceptable and can be screened out with ease.

Remedies:

– Reduce draught by closing off some of the air inlet area at the base of the shaft.

– Reduce the stone size.

– Use more fuel.

A *wide range in quality* is likely to be the most common feature of a small low technology operation. It results from uneven firing which is caused by:

– A wide range in stone feed size;

– Uneven loading of stone and fuel, i.e. bad mixing of stone and fuel;

– Draught being drawn into the kiln unevenly, perhaps due to the wind blowing more into one air supply hole than another;

– Hanging.

Remedies:

– Stricter control of stone feed size and grading.

– Stricter control on charging.

– Poking regularly to prohibit the limestone from hanging in the shaft.

– Designing discharge openings so as to ensure the even drawing of air into the kiln.

Hanging or sticking of the material in the shaft is a result of the lateral pressure on the walls of the kiln of the descending column of material. Hanging can be described as the formation of an arch across the shaft which inhibits the continuous flow of material through the firing zone.

The column of material should be inspected and poked regularly to maintain a consistent flow of material. If arches are allowed to form and the flow inhibited, the material above the arch when poked and thus brought into the firing zone will not have been sufficnetly preheated to allow the core of the stone lumps to reach the dissociation temperature. Thus the core of the material will remain unburnt.

3.4 Hydration

Hydration or slaking can be described as the process of adding a quantity of water to lumps of lime causing them to disintegrate to a powder, putty or limewash. This chemical reaction between lime and water results in the development of a considerable amount of heat (see "chemical reactions of lime", section 1.5). The form into which lime is slaked depends on the use for which it is required. In the case of lime for use in plasters and mortars, which could be either in the form of a putty or a dry powder, all the implications of using either one of the two, must be carefully studied before any decision is made. Qualitatively, the advantage of the use of lime putty over a dry hydrate are that it is likely to contain a greater portion of fine lime particles and will therefore be more plastic, a characteristic which is preferred in mortars and plasters. Also, the product is likely to be more fully slaked and will therefore be less likely to present any of the typical popping and chracking problems that may occur due to the presence of unslaked material. However, more water is required to slake it, so the economic implications of the availability of water take precedence. In a dry area where distances to the market are long, it is likely to be preferable to transport and slake quicklime lumps at sources of water nearer the market than to bring water to the production site and then transport dry lime hydrate or lime putty over a long distance to the market.

3.4.1 Water for hydration

The water used in hydration may be drinkable or even brackish borehole water but water containing a large proportion of organic material can have a bad effect on the lime hydrate. The water required to slake quicklime to:

a) a *dry lime hydrate* is around 550 litres per tonne quicklime,

b) a *lime putty* is around 1300 litres per tonne quicklime depending on the consistency preferred.

The exact quantities required will vary from one quicklime to another and can best be determined by trial and error. In general however, highly reactive porous type quicklime will require a greater proportion of water than dense or overburnt quicklime. Also, dolomitic lime will normally require less water in hydration since only small portions of the MgO content, if any, actually hydrate.

3.4.2 Rate of hydration

The rate of hydration is determined by the type of stone that is fired to start with, and the conditions to which it is subjected during firing. Complete hydration can take place in a matter of a few minutes or continue over a period of months. The factors which determine the rate of hydration are:

a) A quicklime with a high MgO content has a slow rate of hydration since it is normally overburnt when fired at the temperature necessary to calcine $CaCO_3$.

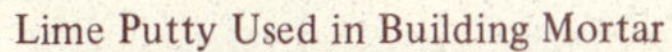

Lime Putty Used in Building Mortar

b) A pure, high calcium lime hydrates faster than one containing impurities. Impurities cause the stone to overburn at lower temperatures which reduces porosity and consequently the rate of hydration.
c) A lightly burnt, porous quicklime will hydrate faster than an overburnt, dense one.
d) If quicklime is crushed to a size smaller than 25 mm the rate of hydration is increased.
e) The rate of hydration increases with an increase of both the temperature of the quicklime lumps and of the water used for slaking. If the quicklime lumps are slaked immediately after they are extracted from the kiln, i.e. whilst they are still slightly hot, and the heat from the waste gases is used to heat the water of hydration, the rate can be increased. With some limes a 10 % increase in water temperature could as much as double the rate of hydration.
f) The use of an excess amount of water applied to the quicklime at a rapid rate retards the rate of hydration.
g) If the quicklime and water mixture is agitated during hydration the rate is increased.

3.4.3 Hydration of magnesian limes

Magnesian limes are slower in slaking than calcium limes due to the sintering caused by the overburning of the magnesium carbonate portion of the stone. Overburnt magnesian limestone or dolomite hydrates very slowly and is just about impossible to hydrate when impure.
Since most of the magnesium oxide portion remains unslaked when using hand slaking methods, less water will be required. Searle suggests that up to 20 % less water will be required for hydration. The wet slaking method described below (handslaking) is a suitable simple means of slaking magnesian limestone quicklimes. The period in the slaking pits can be extended to 1 month.

3.4.4 Methods of slaking

Quicklime can be hydrated either by mechanical means or by hand slaking. For a small scale operation in a situation where employment creation is a priority, labour is plentiful and can be hired at a sufficiently low cost, and the quality of the material fired differs from one batch to the next requiring special attention in slaking, hand slaking is likely to be the most cost effective, and also preferable from the point of view of quality.

3.4.4.1 Hand slaking

One method of hand slaking to produce a dry lime hydrate is as follows:
1. Spread a manageable amount of quicklime on the slaking floor in a layer around 250 mm thick.
2. Spray slowly, using a watering can, approximately 1/3 of the required water over the surface of the layer, and mix. Do the same for the second and third portions. If the correct amount of water is used, after the third spray the hydrated lime powder formed will be slightly damp.
3. Once the watering procedure is completed pile the hydrate in a mound or store in a

silo for a period of 12–24 hours so that the slaking process may be completed.

4. Screen or separate hydrate and then bag.

The method of hand slaking to produce a lime putty is as follows:

1. Spread lime lumps in a shallow slaking pit.
2. Using a watering can, spray all the water necessary to produce the required consistency lime putty whilst stirring continuously using paddles or some other means.
3. Leave in the slaking pit for 12–24 hours to complete slaking.
4. Take lime putty from the pit and screen out core and overburnt lumps above 2 mm diameter.
5. Take slurry to slaking pits in which it is allowed to slake for a period of a week before using. The oversize core and overburnt material can be slaked separately for a longer period in deep slaking pits and then rescreened. (See hydration of magnesian limes, section 3.4.3)

3.4.4.2 Mechanical slaking

Mechanical methods of slaking follow the same process as hand slaking. The difference is that water is added in specifically measured quantities and at predetermined rates to produce either a dry or wet lime, and mixing is done mechanically. For a dry hydrate, the lime can then be screened by a barrel screen or an air separator can be used to separate the oversize core or overburnt material. The screened or separated lime should ideally not contain more than 5 % of material

Hand Operated Lime Putty Production System

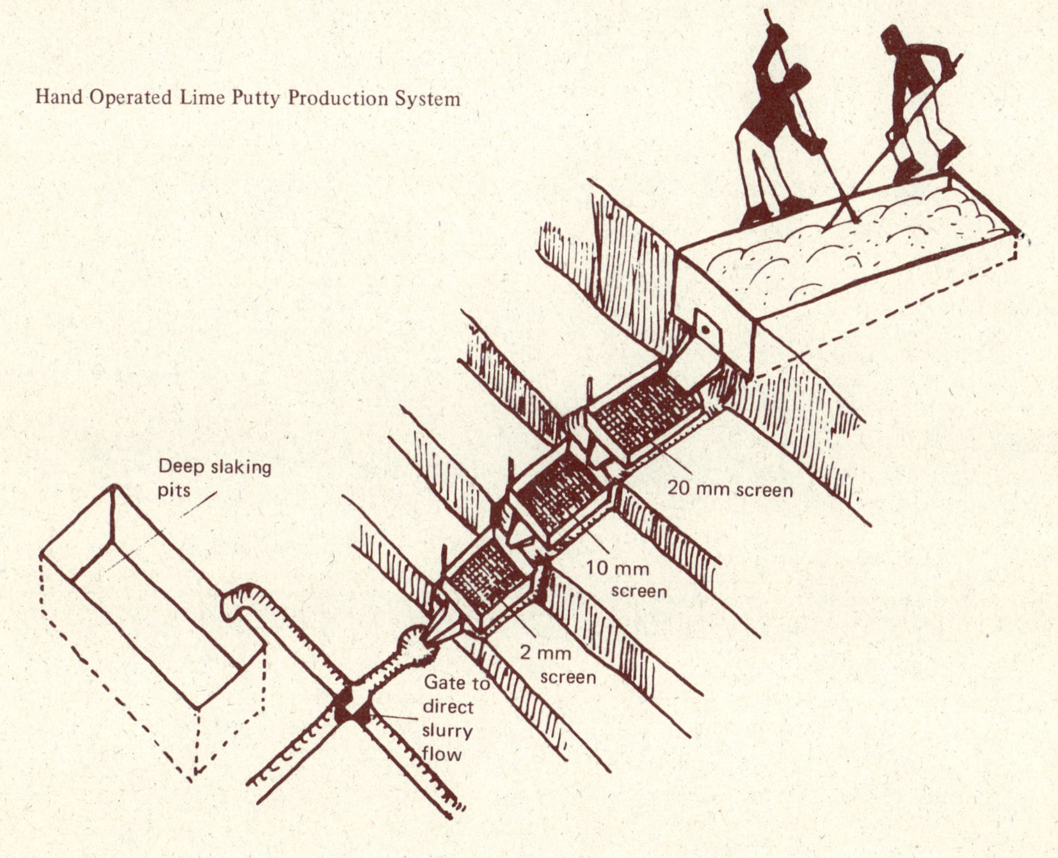

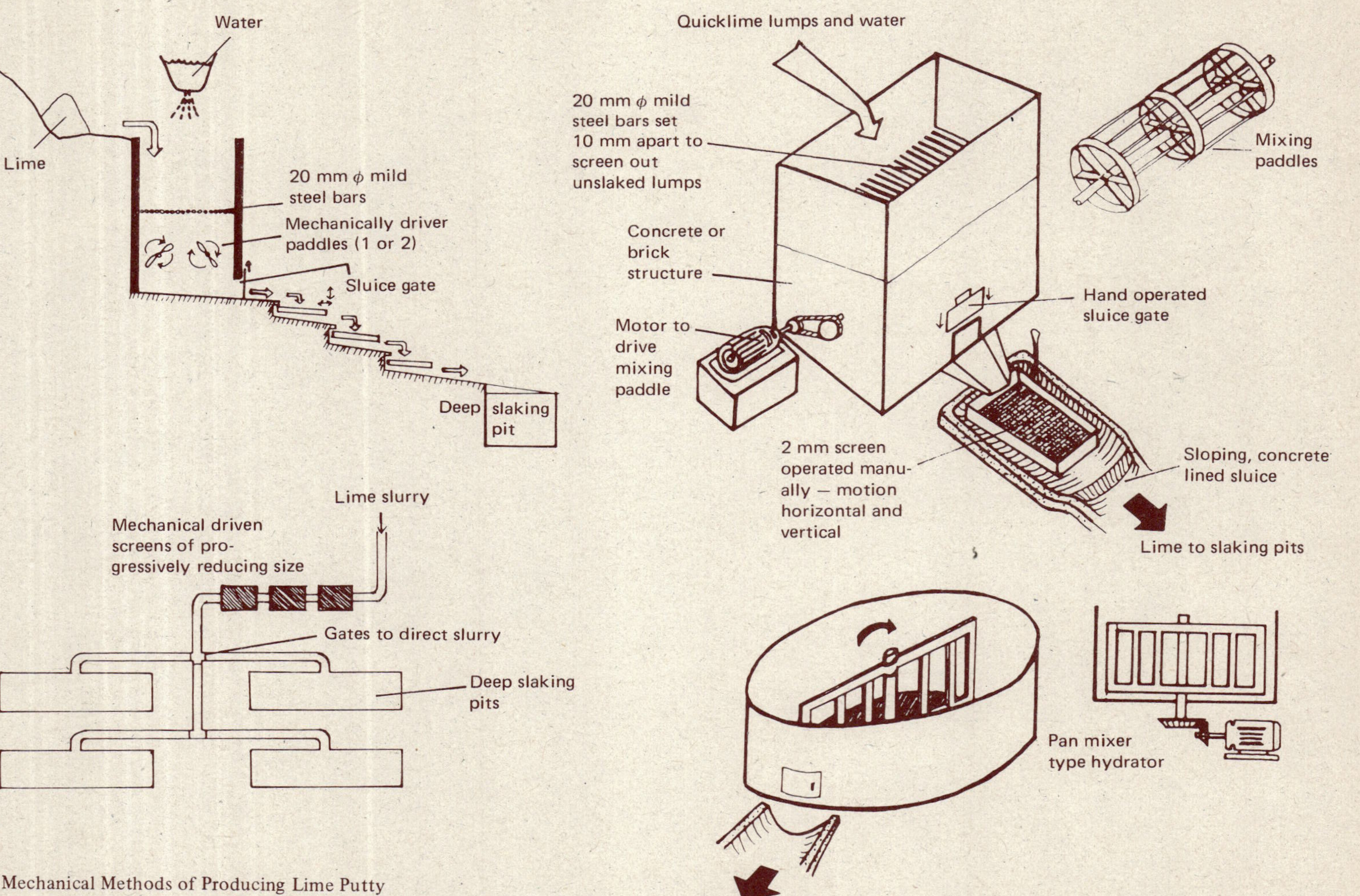

Mechanical Methods of Producing Lime Putty

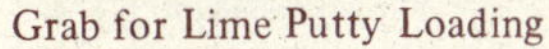

Grab for Lime Putty Loading

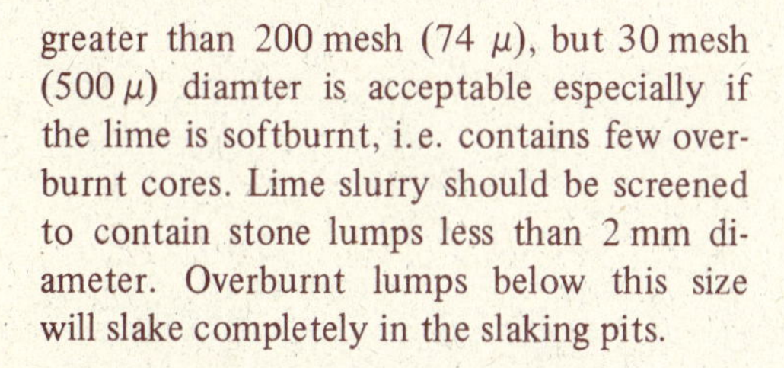

greater than 200 mesh (74 μ), but 30 mesh (500 μ) diamter is acceptable especially if the lime is softburnt, i.e. contains few overburnt cores. Lime slurry should be screened to contain stone lumps less than 2 mm diameter. Overburnt lumps below this size will slake completely in the slaking pits.

3.4.5 Causes and effects of a badly hydrated lime

Drowning: – If water is supplied too quickly to the quicklime lumps, their surfaces will hydrate but the water will tend not to penetrate to the interiors of the lumps. Special care must be taken to avoid this when impure, dense or dolomitic quicklimes are being slaked since they are particularly susceptible to this phenomenon due to their relatively low hydration temperature and rate of hydration. The result of drowning is incomplete hydration.

Burning in hydration occurs when too little water is mixed with the quicklime. A portion of the quicklime remains unslaked.

Whatever the cause, the effect of a badly hydrated lime is the presence of unhydrated cores in the product. The presence of these in lime for use in mortars and plasters is particularly harmful since it is the cause of cracks in mortar joints and popping in plasterwork.

Recarbonation: – The quicklime lumps left lying for extended periods in the open are slaked by absorbing moisture from the air. In addition to slaking however, they absorb free CO_2 from the atmosphere, thus recarbonating. Recarbonation results in a reduction in the available oxide content (CaO and MgO) which reduces the quality of the lime.

4. Testing and Quality Control

The objective of this section is to describe the testing procedures which could be adopted if specialized personnel or facilities are not available at a sufficiently low cost, or when required. However, it must be stressed that to avoid a waste of effort and money, it may be better to wait a little longer for, or spend a little more money on such services. The quality control tests can be conducted adequately on site.

4.1 Geological investigation (the geological surveying)

The search for a suitable deposit should be avoided by the layman. If this is definitely not possible it should be preceded by considerable reading on the geological aspects of limestone, e.g. the forms in which it can be found, its bedding, and its mineralogy. Further, any reports, studies and investigations which relate to the proposed project should be collected and analysed.

Once in the limestone region one has to:
– Establish the average thickness of the overburden.
– Determine the physical nature of the bedding, e.g. if a deposit is calcrete, is it a hardpan type.
– Where the deposit is not thick bedded and uniform at least in appearance, determine *roughly* the relative quality of the different layers by using a solution of hydrochloric acid on samples of the different layers. A drop of acid applied to the rock will cause it to froth and the greater the frothing the greater the available lime in the rock. If a difference cannot be determined, representative samples taken from the different borrow pits and trenches must be laboratory tested, as described later.
– Determine the average thickness of the different layers over the area and thus estimate the average volume of material available.
– Determine the strike and dip of the layer (see the fig. p. 26).

This information can be determined by using the following methods:

– Inspect geographical features of the area, such as river beds or borrow pits, which expose the strata of the deposit. These may not be sufficient, or not present at all, in which case the following methods must be used to acquire information or supplement that available:
– Clear the ground cover (overburden) at random locations over the area to establish its average depth and the ease with which it can be cleared.
– Considering the geography of the area, excavate a series of pits and trenches at suitable relative positions to acquire the geological information as stated above.
– Take representative samples from the different deposits and from different locations in each deposit if necessary, for laboratory testing. Record *very* accurately the sources of the samples, i.e. which pit or trench they came from and the exact position in these.

4.2 Laboratory testing

Laboratory testing of samples taken during the geological survey should, if possible, be carried out by the Geological Survey Department or some other institution with specially trained personnel and the necessary equipment and materials. Analyses to provide the following information should be conducted to enable comparison and selection:

- available oxide content (CaO + MgO),
- expected loss on ignition,
- reactivities of quicklime,
- types and quantities of impurities,
- relative porosity and hardness,
- colour of quicklime.

If it is not possible to use the services of such an institution, the samples could be adequately compared by firing them in the field testing kiln and analysing them using the techniques described below.

4.3 Field testing

A representative sample of limestone from each deposit of sufficient quantity must be prepared for firing in the kiln.

Operation of field testing kiln (see p. 66)

The tests should be conducted in a manner which approximates the real conditions in the full size kiln.

1. Load kiln with layers of fuel and lime, with fuel as the first layer.
2. Start a fire in the fireplace to light the fuel in the bottom layer.
3. Once the bottom fuel layer has caught fire shut off the front of the fireplace to reduce the draught through the kiln.
4. After 36 hours the burning will be completed. Remove the grid rods and extract the quicklime lumps from the kiln.
5. Separate the well burnt lumps from the underfired.
6. Take a representative sample of the well burnt quicklime lumps and execute tests 2 and 3 described below immediately, or pack the lumps in an airtight plastic bag or tin if these tests have to be delayed.
7. Take a further representative sample of well burnt lumps (20 litres) from the remaining portion and conduct test 1.
8. Slake the remaining well burnt quicklime lumps and conduct test 4.

The main advantage of firing in the field test kiln, beside providing samples for laboratory testing at a relatively low cost, is that it enables the project manager to observe the behaviour of the limestone and fuel when fired under field conditions. For example if a limestone decrepitates during firing it immediately renders it unacceptable. Similiarly, if a fuel pollutes the lime excessively, the fuel will be unacceptable. Further, it enables him to determine approximately the necessary design and operating conditions that will be required.

4.4 Quality control

At the level of technology referred to in this text little can be done during firing to make adjustments which affect quality. The only direct control is by making adjustments that will affect the extent of the draught through the shaft, and these changes will be based on experience of the effect of the prevailing winds on the quality of the product. Thermocouples are an indirect method of quality control which serve to warn against overburning.

As Sobek points out, thermocouples merely measure the temperature of the interface between the column of material and the kiln wall. They cannot measure how efficiently the limestone has been calcined. They can only warn against overheating. Thermocouples are particularly useful during the firing trial stage of the project, i.e. when trials in

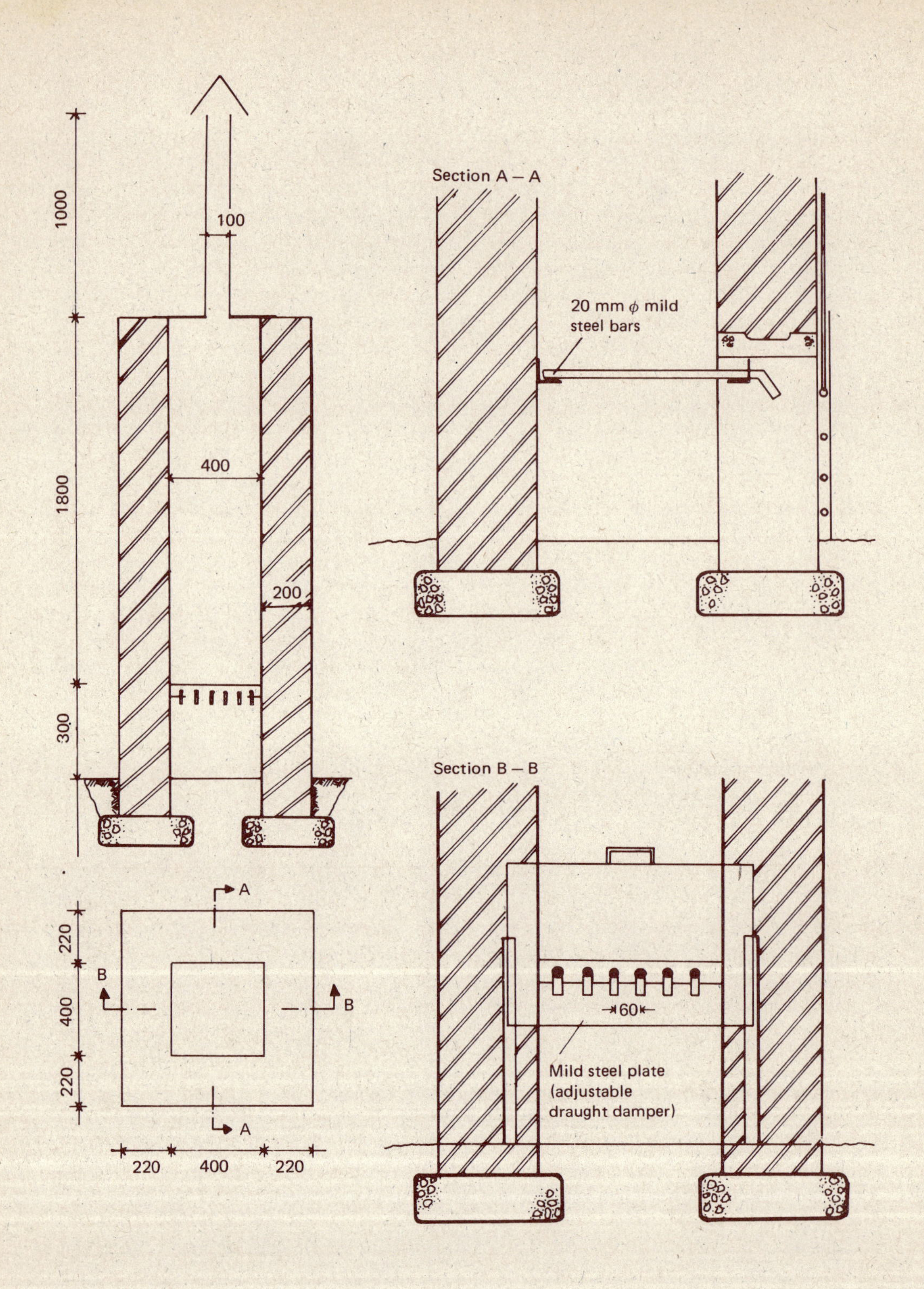

Field Testing Kiln

the full size kiln are being conducted to determine the necessary operating conditions.
Quality control tests must be carried out on the final product and adjustments made according to the results. The quality control tests that should be carried out at regular intervals, i.e. on a daily basis, are 1 and 2 and less regularly, say once a week, are 3 and 5, as explained below.

4.5 Tests

These tests are used both at the testing stage of the project when samples from the various deposits are being tested for quality and compared, and also during production for quality control purposes.

Test 1: Loss on ignition test (LOI)

The LOI test can be conducted at regular intervals during production to monitor the relative degree of calcination. It should be accompanied by a thorough visual inspection. It is also used in the testing stage to compare LOI of limestone from different deposits.

Apparatus:
– Container of fixed volume (20 or 50 litres), such as a bucket.
– Scale of sufficient size to weigh the above volume.

Method:
1. Weigh container (Wb).
2. Weigh the container filled with a representative sample of limestone feed (Wf).
3. Weigh the container filled with a representative sample of quicklime lumps (Wa),
Or
Conduct the above weighing exercise several times (5 will suffice) with different batches of limestone feed and quicklime lumps to determine average figures for (Wf) and (Wa).

4. Calculate the % weight lost on ignition using the following formula:

$$\frac{Wf - Wa}{Wf - Wb} \times 100 = \% \text{ weight lost in ignition } (\% \text{ LOI}).$$

The % LOI can be compared with a standard LOI figure calculated under precise laboratory conditions to establish the relative degree of burning, or if this is not available the theoretical value can be used. The volume of quicklime which has been used in the weighing exercise must be inspected to determine to what degree the limestone is overburnt. If firing is conducted correctly there should be no, or a very little, underburnt stone, but this should also be checked for.

Overburnt quicklime lumps can be distinguished by:
a) A difference in colour compared to lightly burnt lumps.
b) A relative difference in weight between lumps of approximately the same size. Overburnt material will be heavier than lightly burnt material.
c) Shrinkage due to overburning may cause cracks to appear.
d) When tapped lightly with a hammer overburnt lumps will produce a sharp ringing tone compared to the tone produced by a lightly burnt stone.
Underburnt stone can be distinguished by the presence of a hard core which has a different colour and texture to the burnt portion. The amount of overburnt material must be estimated and compared with that normally expected (say around 10 %). A variation in the proportion of overburnt stone from that normally expected will produce a variation in the weight of the quicklime lumps (Wa), which will in turn provide an inconsistent and unreliable % LOI figure. The process of inspecting the volume

visually is in itself a valuable quality control exercise. It allows the operator to inspect the product systematically and at close range.

Test 2: Reactivity assessment of quicklime

The addition of water to quicklime to produce a lime hydrate results in the evolution of heat. A lightly burnt quicklime will evolve heat, i.e. react, at a faster rate than will a hard, overburnt quicklime. This phenomenon is used in this test to monitor the reactivity and hence the degree of burning of the quicklime produced. It is also used in comparing limestones from different deposits for the purpose of selection.

Apparatus:

- Pestle and mortar.
- Nr. 7 mesh sieve (2.83 mm).
- Scale to weigh from 50 g to 500 mg.
- Thermos flask.
- Thermometer reading at least to 100 °C.
- Graph paper.

Method:

1. Take a representative sample of quicklime lumps of around 2 kg and crush them to small fragments.
2. Take a 200 g representative sample from the fragments, pulverize in the pestle and mortar and pass the whole 200 g sample through a 7 mesh sieve.
3. Place 170 ml of water at room temperature in the thermos flask. Weigh out 50 g of the screened material and add it to the water in the thermos flask.
4. Record the rise in temperature of the mixture in the flask at one minute intervals.
5. Continue taking readings for 24 minutes.
6. Plot the temperature-time curve on graph paper and compare with a standard curve or with previous curves.

The previous two tests together with a thorough visual inspection performed on a daily basis, will suffice for monitoring purposes on a small project.

Test 3: Determination of available lime by the RAPID SUGAR TEST

This procedure is one by Boynton, who considers it a simple and accurate test.

Apparatus:

- 300 ml "Erlenmeyer flask" (Conical flask indicating approximate volumes).
- 100 ml burette with a stand.
- Scale weighing 500 to 1000 mg.
- No. 100 mesh sieve.

Materials:

- CO_2 free distilled water.
- Hydrochloric acid (17.5 ml per litre of distilled water).
- Anhydrous sodium carbonate (Na_2CO_3) (0.85 g).
- Methyl orange indicator.
- Sucrose – granulated sugar is satisfactory (15 g).

Method:

1. Take representative sample of hydrated lime and screen through no. 100 mesh sieve.
2. Take a 500 mg sample and brush it into the Erlenmeyer flask containing 20 ml distilled water.
3. Cork the flask, swirl and heat for 2 minutes.
4. Add 150 mg water and 15 g granulated sugar.
5. Re-cork flask and shake at intervals for 5 minutes.
6. Allow to stand for 30 minutes to 1 hour.
7. Add 2 drops phenolphthalein.
8. Wash down sides of flask and stopper with water.

9. Titrate in the original flask with the standard HCl solution (see note below). Add 90 % of the estimated amount of acid solution before shaking the flask and then complete titration with the final 10 % of the acid solution being fed slowly until the pink colour disappears.
10. Note the reading: 1 ml of acid solution is equivalent to 1 % available lime expressed as CaO.

Note: "A standard HCl solution is prepared of 15.7 ml of HCl (sp. gr. 1.18) per litre of CO_2-free distilled water. The solution is standardized against 0.85 g of anhydrous $NaCO_3$ with methyl orange as indicator, so that this amount will neutralize exactly 90 ml of standard HCl solution. In adjusting for this, add more water if it is too strong or more acid if too weak." (Boynton, p. 544)

Test 4: Comparison of lime plasticity (bulk density test)

The plasticity of lime is one of its valuable features when it is to be used in a mortar or plaster. This is due to its great specific surface area ($13000\ cm^2/g$ compared to around $3200\ cm^2/g$ for portland cement), or fineness.
A measure of fineness or specific surface area of a lime hydrate will indicate its plasticity. This can be done by measuring its bulk density.

Apparatus:
– Container.
– Scale weighing at least up to 30 kg.

Method:
1. Weigh container (Wc).
2. Fill container with water and weigh (Ww).
3. Fill container with hydrated lime (powder form) and weigh (Wb). (The different lime hydrates for comparison must be filled and compacted in the container in exactly the same manner).

Since 1 kg water = 1 litre, the weight of water (Ww) is equal to its volume (V_W) ($1\ m^3$ = 1000 litres).

$$\text{Bulk density} = \frac{Wb - Wc}{V_w} \quad \frac{kg}{m^3}$$

Commercial lime has a bulk density of around $575\ kg/m^3$. It is possible to compare the bulk densities of various limes and get an idea of the comparative plasticity. The lower the bulk density, the higher the plasticity. This is a rough test valuable when different limestones are being fired in the field testing kiln for comparison and subsequent selection.

Test 5: Soundness test

The soundness test is a very simple but important test. Its purpose is to determine how effectively the quicklime slakes. Small cores of overburnt material may remain in the lime hydrate. They will slake very slowly. If a lime containing such cores is used in a plaster, at some future time the core will slake in the wall causing the material around it to pop out. Hence the commonly known defect "popping". To avoid this defect the lime hydrate sold must be completely slaked, without any core of overburnt material. This test is used to control the quality of lime produced during the course of production but is of particular value during the trial stage of the project when trials are being conducted to determine the best method of hydration.

Apparatus:
– Flat mixing surface and a saucer.
– Broad bladed knife – e.g. spatula.

Method:

1. Mix hydrated lime into a stiff paste on the mixing surface using the spatula.
2. Fill the saucer with the paste leaving a smooth, flat surface.
3. Store indoors and examine daily for three to four weeks.

If pitting or popping occurs the lime can be considered to be unsound.

5. Technical Account of a 3 Tonne/Day Limeburning Operation in Moshaneng, Botswana

5.1 Background

The purpose of this section is to provide technical and operating information of a specific project experience using a kiln design which has been well tried and tested. It should be noted that the information given will not necessarily apply in other circumstances. The description and recommendations are intended to serve as a stimulus for the fieldworker and as a general guide through the trials procedure.

The original plan was to supply the local building industry with lime for use in mortars and plasters. However, in the course of investigations, it was discovered that a demand for agricultural lime, lime for soil stabilization and a small amount of lime for the local tannery and gold mining operations existed.

The project had two major aims: firstly, to provide a quality product which would substitute a portion of the South African imported product and secondly, to concentrate on labour intensive methods to maximize employment creation. Further, the limitations were to use locally available materials for the construction of the kiln, and to design it so as to orient the operating,

repair and maintenance manpower requirements to the local skills levels.
The location of the site was determined by the fact that already crushed material was available on a tailings dump of an abandoned mining operation. Thus the quarrying and dressing aspects of the projects would be limited to hand selection and transporting to the kiln. Local demand was estimated at 1000–1500 tonnes anually.

Raw materials – dolomite, coal

Dolomite: The tailings dump contained 100–150 thousand cubic metres of usable stone. The dump was variable as to stone size and quality but selection was relatively easy. 20 villagers were employed on a "piece rate" basis to select and transport stone to the kiln.
Chemical analysis of the usable stone was as follows:
Average CaO content 31.00 %,
Average MgO content 21.50 %.
Theoretically this could produce 89.96 % quicklime. This would be an acceptable quality lime for the purposes required.
Coal: Low volatility 22 MJ/kg.

5.2 Production process

Dolomite lumps selected and prepared at the tailings dump were brought to the kiln ready for firing. At regular intervals a 20 litre sample was taken, visually checked for quality and weighed. A 5 % variance in weight from a predetermined mean was set as acceptable. The purpose of this was to control the grading of the dolomite lumps.
The stone feed and coal were then brought to the top of the kiln by means of wheel barrows and loaded in alternate layers.

A layer of firewood was placed at the base of the column of material to assist the ignition of the fire. The fire was lit (with difficulty) and after a period of approximately 24 hours it reached the firing zone. The burnt lumps could then be extracted from the outlet ports. The temperature in the kiln was monitored by means of thermocouples. The quicklime produced was transported by means of wheel barrows to the slaking floor where it was slaked, and then screened by a barrel screen and bagged in 25 kg bags. A labour force of 14 men was employed on a full-time basis and of 20 on a temporary basis.

Stone Collection in Moshaneng, Botswana

Kiln under Construction

Plant and equipment

– Khadi Village Industries (KVI) type, 3 tonne per day vertical shaft kiln.
– 3 thermocouples.
– Wheel barrows, shovels, sledgehammers and extraction forks.
– Slaking floor.
– Hand operated barrel screen with a 5 mm and 30 mesh screen.

Kiln under Construction

5.3 Trials

Following the chemical analysis of the raw materials by the Geological Survey Department and their suggestions, the stone was fired in the on-site testing kiln (see section 4). Positive results were obtained. The construction of the full size kiln and the execution of the trials followed.
The objective of the trials was to determine the best operating conditions under the prevailing circumstances. It must be noted, particularly for this size of project and level of technology that the most appropriate and effective operating conditions can only be determined after continuous trial and error by the operator during the course of production. The first trial runs, conducted as scientifically and systematically as is practically possible will serve as a means of obtaining the basic operating information on which to continue. Extensive trials, prior to going into production, conducted with strict scientific discipline, will not necessarily

3 Ton per Day KVI-Type Kiln Used in Moshaneng

result in a proportional improvement in quality or cost effectiveness. It will almost certainly burden the project with a disproportionately high implementation cost relative to the total investment cost.

The method adopted for the trials was as follows:

1. Representative samples of stone feed and coal were weighed and the volumes charged recorded.
2. Firing commenced and the temperature at the three kiln zones was recorded at half hourly intervals (need only be at hourly or two hourly intervals).
3. Quicklime was extracted and representativ samples weighed.
4. Output was inspected visually and physically.

Trial Kiln Used in Moshaneng, Botswana

5. Output was slaked and a representative sample sent for chemical analysis.
6. Results were considered and appropriate adjustments made.

The trial results must provide the following information:

1. The correct proportion of coal to stone, volume of layers, stone size.
2. The temperature levels required to produce a wellburnt, uniform quality product.
3. Rate of extraction.
4. Water requirement for slaking.
5. On-going quality control requirements.

5.3.1 Observations during firing trials

– In practice, the range of feed size (hand sorted) could not be limited to better than 10 % each side of the mean.
– The stone shape was found not to vary to any noticeable extent.
– The coal size varied from between 5 to 50 mm diameter lumps.
– Strong gusts of wind, as indicated by the thermocouple, caused the temperature to drop back by 25 °C.
– Wind blew in consistently from the west through the western discharge opening.
– Temperature, where it exceeded 950 °C caused an increase of overburnt material with no obvious changes in the amount of core material.
– The fire rose up the shaft at a rate of around 100 mm per hour through a column of material having a stone feed size of approximately 25 mm diameter.
– The material classified as well-burnt had a 10–15 % portion of unburnt core.
– The quicklime lumps became progressively darker from underburnt to overburnt.
– Well-burnt stone weighed less and was softer than overburnt stone.
– The loss on ignition (LOI) was around 28 % for wellburnt material, having a 10 % core.

TRIAL NUMBER	STONE:COAL RATIO	STONE SIZE	TEMPERATURE RANGE	UNDERBURNT	CORE IN UNDERBURNT LUMPS	WELLBURNT	OVERBURNT	COMMENTS
I	4 : 1	125 mm	850-900°C	25%	80%	65%	10%	-Sporadic strong gusts of wind through the western discharge opening.
II	3 :1	100 mm	950-1000°C	15%	70%	70%	15%	-Strong consistent breeze.
III	4 :1	75 mm	950-1000°C	10%	70%	60%	30%	-Light breeze -Compared to previous two firings the temperature rose to the firing zone very slowly.
IV	5 :1	125 mm	800-850°C	30%	80%	60%	10%	-Strong breeze. -Fast rise in temperature up the shaft.
V	4 :1	100 mm	900-950°C	15%	75%	70%	10%	-Inconsistent breeze (Continuous operation)

Trials Results

– The percentage under- and overburnt lime was determined by handsorting a representative sample of output and visually estimating the proportions. Handsorting was based on differences in colour and weight.

– The percentage core was determined by slaking a representative sample of the under burnt stone. A portion of it remained unslaked which constituted the core. The proportions were visually estimated.

– The rate of extraction was between 600–700 litres every 4 hours.
– A large variance existed in the degree of calcination of the quicklime lumps.
– The bulk density of the stone feed was 1430 kg/m^3 and the bulk density of coal 980 kg/m^3.

5.3.2 Batch versus continuous operation

Trials were conducted both on a batch and a continuous basis. It was found that the major advantage of the continuous operation was that it had a greater output per unit time. Consequently the fixed cost per unit output was lower. Other advantages were that a slightly more even quality product resulted, the difficulty and headache of starting a fire was avoided and although it was not possible to gauge exactly to what extent the amount of fuel required could be reduced, i.e. a more thermally efficient operation was possible. It was observed that, in addition to the considerable amount of heat lost in heating up the kiln each time a new batch was started, a substantial amount of heat was lost to the atmosphere when the fire reached the upper level of the kiln. The batch operation is definitely wasteful of heat.

However, if the demand is irregular and unreliable or labour cannot be employed on a continuous basis, e.g. in a subsistence farming environment, a batch kiln may be appropriate. The economic implications of the choice must be carefully defined before a decision is made.

5.4 Hydration

It was decided to hydrate the quicklime to a powder on a concrete slaking floor and then screen it using a barrel screen and bag in 25 kg multilayer paper bags. The quicklime was slaked by being spread in a layer 200 mm thick on the slaking floor and water was sprayed on it. After the initial spray the material was mixed, piled in a mound in the centre, respread and sprayed again. This procedure was continued until the pile no longer absorbed the water i.e. the slaked lime produced (pebbles, sand and powder mixture) was slightly damp. It was found that if one waited a few minutes between each spray and then piled the material up high, the quicklime was broken down to powder more effectively. Further, if it was left to lie two days before screening, a portion of the slow slaking, overburnt quicklime, slaked.

The material was screened to material over 5 mm diameter, between 5 mm and 30 mesh. The amount of material in the minus 30 mesh range was around 50 %, that between 5 mm and 30 mesh around 35 % and the remainder over 5 mm diameter. It was observed that the material in the range 5 mm–30 mesh slaked further in the stock pile and a fair amount of minus 30 mesh lime resulted after screening. The minus 30 mesh slaked lime was bagged in 25 kg bags and stored in a dry area.

5.5 Conclusions and recommendations

5.5.1 Firing

The wide variance in quality seemed the most urgent matter to be dealt with. The irregular supply of air sometimes cooled the fire and sometimes provided too little air for combustion.

To alleviate this problem it is recommended that gates are built at all four discharge openings which can be opened and closed by the operator as required. This will provide a slightly better control of the draught.

Other adjustments which would make for a better distribution of heat are:

Barrel Screen Built for Hand Operation for Screening Dry Lime Hydration; Moshaneng, Botswana

– Reduction in the volumes of the stone and coal charged in each layer from 160 litres stone and 40 litres coal to 80 litres stone and 20 litres coal;
– Making the coal size more uniform and distributing it evenly over each layer of stone. The weight ratio must then be checked to ensure that the necessary adjustment in volume is made so that the proportion of coal used is not increased;
– Keeping a careful control of the quality of stone fed into the kiln. The fine crystal variety is preferable, but more important, the quality fed into the kiln should be consistent.

Another problem was the large amount of underburnt material comprising a considerable proportion of unburnt core. It is recommended that the following measures be taken:
– Reduce the stone size to 100 mm to reduce the draught through the kiln.
– Fire with a stone to coal volume ratio of 4:1. The temperature should be kept at between 900–950 °C.

5.5.2 Hydration

Under the circumstances slaking should be continued for the time being on the slaking floor, screened and bagged as described. The change recommended is to try the production of lime putty from the screened oversize material. Small, trial slaking pits should be built in which the oversize material, with an excess amount of water can be allowed to slake for a period of 2–3 weeks. The over 0.5 mm* material should then be screened out. The screened putty must then be tested chemically, including a soundness test, as described in section 4, to ascertain the extent to which the overburnt core remaining in the lime putty causes popping.

(See the fig. p. 61 for a schematic representation of a hand operated hydration plant that could be used.)

If these tests prove successful, before going into full production of the lime putty, the saleability of lime in such a form should be established and the economic implications calculated and evaluated.

* A high calcium lime will not need to be screened through such a fine screen, especially if lightly burnt, but in this instance where the quicklime is both from a dolomitic limestone and contains overburnt material, a 0.5 mm screen is preferable.

Appendix

Useful addresses

African Regional Centre for Technology, BP 2435, Dakar, Senegal (a joint project of UNECA and Organization of African Unity)
Aid Resources Report, Science/Technology Bureau, USAID, Washington, DC 20523, USA
Asian Institute of Technology, PO Box 2754, Bangkok, Thailand
ATOL, Aangepaste Technologie Ontwikkelinglanden, Blijde Inkomststraat 9, 3000 Leuven, Belgium
BRACE Research Institute, Macdonald College of McGill University, Ste Anne de Bellevue, Québec H9X ICO, Canada
BRE. Building Research Establishment, (Overseas Division), Garston, Watford, WD2 7JR, UK
Building & Roads Research Institute (B.R.R.I.), P.O. Box 40, U.S.T., Kumasi, Ghana, West Africa
Cater, Centro Andino de Tecnologia Rural, Casilla 399, Loja, Ecuador
CEMAT, Centro Mesoamericano de Estudios sobre Tecnologia Apropiada, Apartado Postal 1160, Guatemala Ciudad, Guatemala
Centre de Construction et du Logemont (CCL), B.P. 1762, Coccavelli/Lome – Togo
CETAL, Centro de Estudios en Tecnología Apropiada, para América Latina, Casilla 197–V, Valparaiso, Chile
CORT, Consortium on Rural Technology, A – 89 Madhuvan, New Delhi 110092, India
COTA, Collectif d'Echangé pour la Technologie Appropriée, 18 rue de la Sablonnière, 1000 Bruxelles, Belgique
Department of Public Works, P.O. Box 1108, Boroko/Papua-New Guinea
GATE, German Appropriate Technology Exchange, GTZ-GmbH, Postfach 5180, D-6236 Eschborn 1, West Germany
GRET, Group de Recherches et d'Echanges Technologiques, 30 rue de Charonne, 75011 Paris, France
Housing Research and Development Unit (HRDU), University of Nairobi, P.O. Box 30197, Nairobi/Kenya
ITIS, Intermediate Technology Industrial Services, Myson House, Railway Terrace, Rugby CV21 3HT England
NAS, National Academy of Sciences' National Research Council, 2101 Constitution Ave, Washington, DC 24018, USA
Oxfam, 274 Banbury Rd. Oxford OX2 7DZ, UK
PCATT, Phillipine Center for Appropriate Training and Technology, Manila Suite, 1416 F Felipe Agoncillo St, Ermita, Metro Manila, Philippines 2801
Regional Centre for Technology Transfer, P.O. Box 115, Bangalore 560 052, India
Resource Use Institute, Dunmore, Pitlochry, Perthshire PH 16 5EH U.K.
SKAT, Swiss Center for Appropriate Technology, Varnbüelstrasse 14, 9000 St. Gallen, Switzerland
SEDCO, Small Enterprises Development Corporation, P.O. Box 451, Mbabane, Swaziland
T.C.C., Technology Consultancy Centre, University Post Office, U.S.T., Kumasi, Ghana, West Africa
TOOL, Technische Ontwikkeling Ontwikkelingslanden, Mauritskade 61a, 1092 AD Amsterdam, Netherlands
Tranet, PO Box 567, Rangeley, ME 04970, USA
Transport and Road Research Laboratory, (Overseas Unit, Department of Environment), Crowthorne, Berkshire RG 11 6 AU
UNIDO, United Nations Industrial Development Organization, PO Box 300 – A, Vienna, Austria (National Office)
VITA, Volunteers in Technical Assistance, 1815 N Lynn Street, Arlington, VA 22209–2709, USA
VTU, Village Technology Unit, c/o Commissioner for Social Services, PO Box 30276, Nairobi, Kenya

Lime kiln designs with 3.5 and 10 tonnes per day output are available from:
Directorate of the Lime Manufacturing Industry, Khadi Village Industries Commission, 3 Irla Rd, Vile Parle (West), Bombay – 56 (AS), India

Specification for lime:
Limes for building purposes:
American Society for Testing Materials – C6–49 (1968); C206–49 (1968); C207–49 (1968)
British Standards Institution – BS 890–1972
West German Standards Board – DIN 1060–1967
Indian Standard – IS 712–1964

References

Bessey, G.E., Production and use of lime in developing countries. Overseas Building Note Nr. 161. Building Research Station. Garston, Watford, 1975

Boynton, R.S., Chemistry and technology of lime and limestone. John Wiley and Sons Inc. New York, 1980

Ellis, C.I., Village scale production of lime for road construction in Ghana. Transport and Road Research Laboratory. Crowthorne, Berkshire, 1974

Gwosdz, W., Report on quicklime burning tests, Botswana. Geological survey report, WG/33/81. Botswana 1981

Hodges, J.W., The small scale production of hydrated lime. Unpublished report. Road Research Laboratory, Crowthorne, Berkshire

Hosking, J.S., Limestone and lime in the Territory of Papua and New Guinea. Technical Paper No. 21. Division of Building Research, CSIRO, Melbourne 1967

Pollard, A.E., Notes on the lime-kiln constructed at San Antonio quarry Belize, by Sir William Halcrow and partner. Transport and Road Research Laboratory Working paper No. 14. Transport and Road Research Laboratory. Crowthorne, Berkshire, 1977

Sauni, J.T.M. and *Sakula, J.H.*, Oldonyo Sambu pozzolime industry. History, operation and development. Handing over report. Small Industries Development Organization. Arusha, June 1980

Searle, A., Limestone and its products. E. Benn Ltd. London, 1935

Sobek, F., Manufacturing guide. Lime industry. United Nations Industrial Development Organization. May 1975

Spence, R., Alternative cements in India. Intermediate Technology Development Group. London, 1976

Stutterheim, N. Webb, T.L. and *Uranovsky, B.*, Developments in research on the burning and hydration of lime and on its use in building. National Building Research Institute, Council for Scientific and Industrial Research, South Africa. Building Research Congress, 1951